The New Gold Rush

Joseph N. Pelton

The New Gold Rush

The Riches of Space Beckon!

Copernicus Books is a brand of Springer

Joseph N. Pelton
Executive Board, International Association
  for the Advancement of Space Safety
Arlington, VA, USA

ISBN 978-3-319-39272-1   ISBN 978-3-319-39273-8   (eBook)
DOI 10.1007/978-3-319-39273-8

Library of Congress Control Number: 2016957291

© Springer International Publishing Switzerland 2017
This work is subject to copyright. All rights are reserved by the Publisher, whether the whole or part of the material is concerned, specifically the rights of translation, reprinting, reuse of illustrations, recitation, broadcasting, reproduction on microfilms or in any other physical way, and transmission or information storage and retrieval, electronic adaptation, computer software, or by similar or dissimilar methodology now known or hereafter developed.
The use of general descriptive names, registered names, trademarks, service marks, etc. in this publication does not imply, even in the absence of a specific statement, that such names are exempt from the relevant protective laws and regulations and therefore free for general use.
The publisher, the authors and the editors are safe to assume that the advice and information in this book are believed to be true and accurate at the date of publication. Neither the publisher nor the authors or the editors give a warranty, express or implied, with respect to the material contained herein or for any errors or omissions that may have been made.

Printed on acid-free paper

This Copernicus imprint is published by Springer Nature
The registered company is Springer International Publishing AG
The registered company address is: Gewerbestrasse 11, 6330 Cham, Switzerland

This book is dedicated to all of the young people who aspire to be involved in the gold rush in the skies, who are now studying outer space around the world. These include students at the International Space University of Strasbourg, France, the McGill University Centre for Research in Air and Space Law in Montreal, Canada, the Leuven Centre for Global Space Governance in Belgium, the Space Policy Institute at George Washington University, the University of Capetown in South Africa, and the many other universities and institutes offering courses in space policy and law, space transportation, and space science and applications.

Preface

This book grew out of something called the Montreal Declaration. This short declaration was unanimously adopted by an international group of about a hundred space scientists, engineers, and lawyers concerned with the future development and governance of outer space in a time of some entrepreneurial innovation, global change, and some would even say turmoil. It called for an interdisciplinary investigation of all of the elements of change in the world of satellite applications and space exploration in order to assess what was new and revolutionary on the space horizon and what new forms of governance might be needed.

This was not an attempt to reject or turn back the forces of change. Rather it was a call for the study of the innovations that would give rise to a new era of space activities and to see what innovations in the international regulatory and space governance regime might help unlock the potential of the future without giving rise to conflicts in space. Heaven knows there are plenty of conflicts right here on planet Earth. Some elements of change are clear. There are more and more corporate activities in space, and space law is essentially aimed at nations and not industrial enterprises.

There are today a number of new and developing space enterprises and activities that include space mining, the installation of solar power satellites, on-orbit servicing and retrofitting of satellites, and attempts to cope with the problem of orbital debris—including active removal, or the recycling of space junk in the skies. There are new military and defense-related capabilities in the skies, and some of these relate to the idea of planetary defense, which means the deployment of technologies in the skies to detect and monitor cosmic hazards such as asteroids, comets, and solar storms as well as systems to actually defend Earth against these perils from outer space.

The result of this 2-year-long effort is a book entitled *Global Governance of Outer Space*. Space scientists and lawyers will undoubtedly find a book on such a topic to be fascinating, but the general public—perhaps not so much.

However, the general public really has a vested interest in knowing about the practical opportunities represented by what is called "New Space." In this New Space world there are new jobs, new wealth, new opportunity, and new potential conflicts among nations.

It is this practical knowledge about the future of space that this book is all about. We have sought to explain in simple language without technical formulas or arcane rules of space law what John Q. Public—or Jill X. Public—needs to know that is relevant to future job opportunities, totally new types of space industries, as well as truly serious space hazards that could have a devastating impact on our lives if we don't take the right protective steps. The bottom line is that outer space is relevant to the lives of modern men, women, and children in ways that were never true in the past.

In short, there are changing opportunities, new corporate activities in space, new sources of wealth, and even new sources of disputes that could lead to conflict over the future of space.

The New Space industry leaders may not be who you think they are. The new operatives in the commercial space game are organizations such as Google, Facebook, and the Tesla-SpaceX complex (within the empire of Elon Musk). Indeed this New Space push is fueled by who we call the space billionaires. At the head of the space billionaire pack are Jeff Bezos, founder of Amazon.com; Paul Allen, co-founder of Microsoft; Elon Musk (founder of Space X, Paypal, and Tesla); Robert Bigelow, owner of Budget Suites; Sir Richard Branson, head of Virgin Galactic; Mark Zuckerberg, founder of Facebook; and electronic game inventor John Carmack, who created "Doom" and "Quake." It is these people that are upending the world of technology and global enterprise at planetary levels who will be prominent in the space business during the twenty-first century.

This book is intended to reveal to a broader audience the cornucopia of new enterprises that could be opening up in the next few years. That future may well include clean energy beamed from space 24 h a day. Or it could mean new economies in space services from new types of communication satellites, remote-sensing companies, or other new types of space enterprises. It could well mean robotic mining of asteroids rich in platinum and rare Earth metals. It could mean solar space shields to protect vital Earth infrastructure as the Van Allen belts lose their protective power due to the shift of the magnetic poles. Most profoundly it might mean the establishment of permanent colonies inhabited by smart robots and humans on both the Moon and Mars. It could mean the start of a whole new era for humans living on different worlds.

This change is driven by new technology, new instruments of military defense and weaponry, new entrepreneurial space enterprises, and a new awareness that there is a need for the sustainability of space just as there is a concern here on Earth with climate change and the sustainability of our terrestrial world. In short the space revolution is part of the overall change that is the twenty-first century. It is closely tied to a future inhabited by smart robots, which will require a redefining of jobs, employment, and wealth. Indeed, it is all tied into the practical meaning of sustainability and the very future of the human race, whether we will survive as a species.

Ignore this book at your own peril. The future is filled with both considerable risk and enormous opportunity.

Washington, DC
August 2016

Joseph N. Pelton
Former Dean, International Space University
Author of *MegaCrunch: Ten Survival Strategies for the 21st Century*

Contents

1 Why This Gold Rush Is Different   1

2 A Space Cornucopia of Jobs, Resources and More   21

3 The Expanding Use of Space in Communications, Navigation,
   Remote Sensing and Weather Satellites   39

4 Commercial Space Transport, On-Orbit Servicing
   and Manufacturing   69

5 Solar Power Satellites and Space Mining   91

6 Space Security, Defense and Weapons   109

7 Protecting Earth from Space Junk, Cosmic Hazards
   and Climate Change   127

8 Space Habitats, Space Colonies and the New Space Economy   141

9	Governing the New Space Economy	159
10	Policing the Gold Rush in the Skies	177
11	Looking Toward a More Hopeful Global Society	191

Appendix: Current Status of the U. S. Commercial Space Launch Competitiveness Act, Public Law 114-90, as of June 2016	213
Glossary of Key Terms and Phrases	221
Index	227

About the Author

Joseph N. Pelton, Ph.D., is the former Dean and Chairman of the Board of Trustees of the International Space University. He also is the Founder of the Arthur C. Clarke Foundation and the founding President of the Society of Satellite Professionals International. Dr. Pelton currently serves on the Executive Board of the International Association for the Advancement of Space Safety. He is the Director Emeritus of the Space and Advanced Communications Research Institute (SACRI) at George Washington University where he also served as Director of the Accelerated Master's Program in Telecommunications and Computers from 1998 to 2004. Previously he headed the Interdisciplinary Telecommunications Program at the University of Colorado-Boulder. Dr. Pelton has also served as President of the International Space Safety Foundation and President of the Global Legal Information Network (GLIN).

Dr. Pelton has been speaker on national media in the USA (PBS New Hour, Public Radio's All Things Considered, ABC, and CBS) and internationally on BBC, CBC, and FR-3. He has spoken before Congress, the United Nations, and delivered talks in over 40 countries around the world. His honors include the Sir Arthur Clarke International Achievement Award of the British Interplanetary Society, the Arthur C. Clarke Foundation Award, the ICA Educator's award, the ISCe Excellence in Education Award, and being elected to the International Academy of Astronautics.

Dr. Pelton is a member of the SSPI Hall of Fame, Fellow of the IAASS, and Associate Fellow of the AIAA. Pelton is a widely published author with some 40 books written, co-authored, or co-edited. His *Global Talk* won the Eugene

Emme Literature Award and was nominated for a Pulitzer Prize. Currently he is co-Editor of *The Global Governance of Outer Space* which is the global study in which over 80 scientific and legal scholars are participating in response to the Montreal Declaration of 2014.

During his career he also held various positions at Intelsat and Comsat including serving as Director of Project SHARE and Director of Strategic Policy for Intelsat. Intelsat's Project SHARE gave birth to the Chinese National TV University that now is the world's largest tele-education program. He received his degrees from the University of Tulsa, New York University, and his doctorate from Georgetown University.

1

Why This Gold Rush Is Different

> *"Ships and sails proper for the heavenly air should be fashioned. Then there will also be people, who do not shrink from the dreary vastness of space."*
> —Johannes Kepler, Letter to Galileo Galilei, 1609

> *"In spite of the opinions of certain narrow-minded people, who would shut up the human race upon this globe, as within some magic circle which it must never outstep, we shall 1 day travel to the moon, the planets, and the stars, with the same facility, rapidity, and certainty as we now make the voyage from Liverpool to New York!"*
> —Jules Verne, From the Earth to the Moon, 1865

> *"The choice, as Wells once said, is the Universe—or nothing… The challenge of the great spaces between the worlds is a stupendous one; but if we fail to meet it, the story of our race will be drawing to its close. Humanity will have turned its back upon the still untrodden heights and will be descending again the long slope that stretches, across a thousand million years of time, down to the shores of the primeval sea."*
> —Arthur C. Clarke, last words of his first book, Interplanetary Flight, 1950

Are We Humans Doomed to Extinction?

What will we do when Earth's resources are used up by humanity?

The world is now hugely over populated, with billions and billions crammed into our overcrowded cities. By 2050, we may be 9 billion strong, and by 2100 well over 11 billion people on Planet Earth. Some at the United Nations say we might even be an amazing 12 billion crawling around this small globe. And over 80% of us will be living in congested cities. These cities will be ever more vulnerable to terrorist attack, natural disaster, and other plights that come with overcrowding and a dearth of jobs that will be fueled by rapid automation and the rise of artificial intelligence across the global economy. We are already rapidly running out of water and minerals. Climate change is threatening our very existence. Political leaders and even the Pope have cautioned us against inaction. Perhaps the naysayers are right. All humanity is at tremendous risk. Is there no hope for the future?

This book is about hope. We think that there is literally heavenly hope for humanity. But we are not talking here about divine intervention. We are envisioning a new space economy that recognizes that there is more water in the skies that all our oceans. There is a new wealth of natural resources and clean energy in the reaches of outer space—more than most of us could ever dream possible.

There are those that say why waste money on outer space when we have severe problems here at home? Going into space is not a waste of money. It is our future. It is our hope for new jobs and resources. The great challenge of our times is to reverse public thinking to see space not as a resource drain but as the doorway to opportunity. The new space frontier can literally open up a "gold rush in the skies."

In brief, we think there is new hope for humanity. We see a new a pathway to the future via new ventures in space.

For too long, space programs have been seen as a money pit. In the process, we have overlooked the great abundance available to us in the skies above. It is important to recognize there is already the beginning of a new gold rush in space—a pathway to astral abundance. "New Space" is a term increasingly used to describe radical new commercial space initiatives—many of which have come from Silicon Valley and often with backing from the group of entrepreneurs known popularly as the "space billionaires." New space is revolutionizing the space industry with lower cost space transportation and space systems that represent significant cost savings and new technological breakthroughs. "New Commercial Space" and the "New Space Economy" represent more than a new way of looking at outer space. These new pathways to the stars could prove vital to human survival.

If one does not believe in spending money to probe the mysteries of the universe then perhaps we can try what might be called "calibrated greed" on for size. One only needs to go to a cubesat workshop, or to Silicon Valley or one of many conferences like the "Disrupt Space" event in Bremen, Germany, held in April 2016 to recognize that entrepreneurial New Space initiatives are changing everything [1].

In fact, the very nature and dimensions of what outer space activities are today have changed forever. It is no longer your grandfather's concept of outer space that was once dominated by the big national space agencies. The entrepreneurs are taking over.

The hopeful statements in this book and the hard economic and technical data that backs them up are more than a minority opinion. It is a topic of growing interest at the World Economic Forum, where business and political heavyweights meet in Davos, Switzerland, to discuss how to stimulate new patterns of global economic growth.

It is even the growing view of a group that call themselves "space ethicists." Here is how Christopher J. Newman, at the University of Sunderland in the United Kingdom has put it:

> *Space ethicists have offered the view that space exploration is not only desirable; it is a duty that we, as a species, must undertake in order to secure the survival of humanity over the longer term. Expanding both the resource base and, eventually, the habitats available for humanity means that any expenditure on space exploration, far from being viewed as frivolous, can legitimately be rationalized as an ethical investment choice.* (Newman)

On the other hand there are space ethicists and space exobiologists who argue that humans have created ecological ruin on the planet—and now space debris is starting to pollute space. These countervailing thoughts by the "no growth" camp of space ethicists say we have no right to colonize other planets or to mine the Moon and asteroids—or at least no right to do so until we can prove we can sustain life here on Earth for the longer term.

However, for most who are planning for the new space economy the opinion of space philosophers doesn't really float their boat. Legislators, bankers, and aspiring space entrepreneurs are far more interested in the views of the super-rich capitalists called the space billionaires.

A number of these billionaires and space executives have already put some very serious money into enterprises intent on creating a new pathway to the stars. No less than five billionaires with established space ventures—Elon Musk, Paul Allen, Jeff Bezos, Sir Richard Branson, and Robert Bigelow—have invested millions if not billions of dollars into commercializing space. They are developing new technologies and establishing space enterprises that can bring the wealth of outer space down to Earth. This is not a pipe dream, but will increasingly be the economic reality of the 2020s.

These wealthy space entrepreneurs see major new economic opportunities. To them space represents the last great frontier for enterprising pioneers. Thus they see an ever-expanding space frontier that offers opportunities in low-cost space transportation, satellite solar power satellites to produce clean energy 24 h a day, space mining, space manufacturing and production, and eventually space habitats and colonies as a trajectory to a better human future. Some even more visionary thinkers envision the possibility of terraforming Mars, or creating new structures in space to protect our planet from cosmic hazards and even raising Earth's orbit to escape the rising heat levels of the Sun in millennia to come.

Some, of course, will say this is sci-fi hogwash. It can't be done. We say that this is what people would have said in 1900 about airplanes, rocket ships, cell phones and nuclear devices. The skeptics laughed at Columbus and his plan to sail across the oceans to discover new worlds. When Thomas Jefferson bought the Louisiana Purchase from France or Seward bought Alaska, there were plenty of naysayers that said such investment in the unknown was an extravagant waste of money. A healthy skepticism is useful and can play a role in economic and business success.

Before one dismisses the idea of an impending major new space economy and a new gold rush, it might useful to see what has already transpired in space development in just the past five decades. The world's first geosynchronous communications satellite had a throughput capability of about 500 kb/s. In contrast, today's state of the art Viasat 2—a half century later—has an impressive throughput of some 140 Gb/s. This means that the relative throughput is nearly 300,000 greater, while its lifetime is some ten times longer (Figs. 1.1 and 1.2).

Each new generation of communications satellite has had more power, better antenna systems, improved pointing and stabilization, and an extended lifetime.

Fig. 1.1 Syncom 2 launched in 1963 with an equivalent throughput of about 500 kb/s (Image courtesy of the Comsat Legacy project.)

1 Why This Gold Rush Is Different 5

Fig. 1.2 The Viasat 2 with a remarkable throughput 300,000 greater than Syncom (Image courtesy of Via Satellite.)

And the capabilities represented by remote sensing satellites, meteorological satellites, and navigation and timing satellites have also expanded their capabilities and performance in an impressive manner. When satellite applications first started, the market was measured in millions of dollars. Today commercial satellite services exceed a quarter of a billion dollars. Vital services such as the Internet, aircraft traffic control and management, international banking, search and rescue and much, much more depend on application satellites. Those that would doubt the importance of satellites to the global economy might wish to view on You Tube the video "If There Were a Day Without Satellites?" [2].

Let's check in on what some of those very rich and smart guys think about the new space economy and its potential. (We are sorry to say that so far there are no female space billionaires, but surely this, too, will come someday soon.)

Of course this twenty-first century breakthrough that we call the New Space economy will not come just from new space commerce. It will also come from the amazing new technologies here on Earth. Vital new terrestrial

The Visionaries Leading the Charge into New Space Enterprises

On creating "a Million Person Colony" on Mars: "I want to make rockets 100 times, if not 1000 times better. The ultimate objective is to make humanity a multi-planet species. Thirty years from now, there'll be a base on the Moon and on Mars, and you would need a million people to be going back and forth on SpaceX rockets...to recreate the entire industrial base on Mars...people to mine and refine all of these different materials, in a much more difficult environment than Earth. There would be no trees growing. There would be no oxygen or nitrogen that are just—there. No oil."(**Elon Musk,** president of SpaceX and Tesla.)

On his space business, Virgin Galactic: We'll go into orbit. We'll go to the Moon. This business has no limits. (**Richard Branson,** reported in *Wired* magazine January 2005.)

On why space is the next frontier: What should exist? To me, that's the most exciting question imaginable. What do we need that we don't have? How can we realize our potential? As a species, we've always been discoverers and adventurers, and space and the deep ocean are some of the last frontiers. (**Paul Allen,** co-founder of MicroSoft, in "brainy quotes".)

On change: "Here's to the crazy ones, the misfits, the rebels, the troublemakers, the round pegs in the square hole—they're not fond of rules... You can quote them, disagree with them, glorify or vilify them, but the only thing you can't do is ignore them because the ones who are crazy enough to think that they can change the world are the ones who do." (**Steve Jobs,** founder of Apple.)

On investing $275 million in New Space: "We seek to assist human exploration and the discovery of beneficial resources, whether in Low Earth Orbit (LEO), on the moon, in deep space or on Mars". (**Robert Bigelow, CEO of Budget Suites and Bigelow Aerospace.**)

On a space elevator providing low-cost access to space: "It's a phenomenal enabling technology that would open up our Solar System to humankind. It will be robotic, and then 10–15 years after that we'll have six to eight elevators that are safe enough to carry people." (**Peter Swan,** lead author of the International Academy of Astronautics (IAA) report on space elevators.)

On the "can do" spirit. "The "let's just go and do it" mentality will help us finally get off the planet and irreversibly open the space frontier. The capital and tools are finally being placed into the hands of those willing to risk, willing to fail, willing to follow the dreams." (**Dr. Peter H. Diamandis,** chairman of the X-Prize Foundation and CEO of the company Planetary Resources.)

technologies will accompany this cosmic journey into tomorrow. Information technology, robotics, artificial intelligence and commercial space travel systems have now set us on a course to allow us humans to harvest the amazing riches in the skies—new natural resources, new energy, and even totally new ways of looking at the purpose of human existence. If we pursue this course steadfastly, it can be the beginning of a New Space renaissance. But if we don't seek to realize our ultimate destiny in space, *Homo sapiens* can end up in the dustbin of history—just like literally millions of already failed species. In each and every one of the five mass extinction events that have occurred over the last 1.5 billion years on Earth, some 50–80 % of all species have gone the way of the T. Rex, the woolly mammoth, and the Dodo bird along with extinct ferns, grasses and cacti.

On the other hand, the best days of the human race could be just beginning.

If we are smart about how we go about discovering and using these riches in the skies and applying the best of our new technologies, it could be the start of a new beginning for humanity. Konstantin Tsiokovsky, the Russian astronautics pioneer, who first conceived of practical designs for spaceships, famously said: "A planet is the cradle of mankind, but one cannot live in a cradle forever." Well before Tsiokovsky another genius, Leonardo da Vinci, said, quite poetically: "Once you have tasted flight, you will forever walk the earth with your eyes turned skyward, for there you have been, and there you will always long to return."

The founder of the X-Prize and of Planetary Resources, Inc., Dr. Peter Diamandis, has much more brashly said much the same thing in quite different words when he said: "The meek shall inherit the Earth. The rest of us will go to Mars."

The New Space Billionaires

Peter Diamandis is not alone in his thinking. From the list of "visionaries" quoted earlier, Elon Musk, the founder of SpaceX; Sir Richard Branson, the founder of Virgin Galactic; and Paul Allen, the co-founder of Microsoft and the man who financed SpaceShipOne, the world's first successful spaceplane have all said the future will include a vibrant new space economy. They, and others, have said that we can, we should and we soon shall go into space and realize the bounty that it can offer to us.

The New Space enterprise is today indeed being led by those so-called space billionaires, who have an exciting vision of the future. They and others in the commercial space economy believe that the exploitation of outer space may open up a new golden age of astral abundance. They see outer space as a new frontier that can be a great source of new materials, energy and various forms of new wealth that might even save us from excesses of the past.

This gold rush in the skies represents a new beginning. We are not talking about expensive new space ventures funded by NASA or other space agencies in Europe, Japan, China or India. No, these efforts which we and others call

New Space are today being forged by imaginative and resourceful commercial entrepreneurs. These twenty-first century visionaries have the fortitude and zeal to look to the abundance above. New breakthroughs in technology and New Space enterprises may be able to create an "astral life raft" for humanity.

Just as Columbus and the Vikings had the imaginative drive that led them to discover the riches of a new world, we now have a cadre of space billionaires that are now leading us into this New Space era of tomorrow. These bold leaders, such as Paul Allen and Sir Richard Branson, plus other space entrepreneurs including Jeff Bezos of Amazon and Blue Origin, and Robert Bigelow, Chairman of Budget Suites and Bigelow Aerospace, not only dream of their future in the space industry but also have billions of dollars in assets. These are the bright stars of an entirely new industry that are leading us into the age of New Space commerce.

These space billionaires, each in their own way, are proponents of a new age of astral abundance. Each of them is launching new commercial space industries. They are literally transforming our vision of tomorrow. These new types of entrepreneurial aerospace companies—the New Space enterprises—give new hope and new promise of transforming our world as we know it today.

The New Space Frontier

What happens in space in the next few decades, plus corresponding new information technologies and advanced robotics, will change our world forever. These changes will redefine wealth, change our views of work and employment and upend almost everything we think we know about economics, wealth, jobs, and politics. These changes are about truly disruptive technologies of the most fundamental kinds. If you thought the Internet, smart phones, and spandex were disruptive technologies, just hang on. You have not seen anything yet.

In short, if you want to understand a transition more fundamental than the changes brought to the twentieth century world by computers, communications and the Internet, then read this book. There are truly riches in the skies. Near-Earth asteroids largely composed of platinum and rare earth metals have an incredible value. Helium-3 isotopes accessible in outer space could provide clean and abundant energy. There is far more water in outer space than is in our oceans.

In the pages that follow we will explain the potential for a cosmic shift in our global economy, our ecology, and our commercial and legal systems. These can take place by the end of this century. And if these changes do not take place we will be in trouble. Our conventional petro-chemical energy systems will fail us economically and eventually blanket us with a hydrocarbon haze of smog that will threaten our health and our very survival. Our rare precious metals that we need for modern electronic appliances will skyrocket in price, and the struggle between "haves" and "have nots" will grow increasingly ugly. A lack of affordable and readily available water, natural resources, food, health care and medical supplies, plus systematic threats to urban security and systemic warfare are the alternatives to astral abundance.

The choices between astral abundance and a downward spiral in global standards of living are stark. Within the next few decades these problems will be increasingly real. By then the world may almost be begging for new, out-of-the-box thinking. International peace and security will be an indispensable prerequisite for exploitation of astral abundance, as will good government for all. No one nation can be rich and secure when everyone else is poor and insecure. In short, global space security and strategic space defense, mediated by global space agreements, are part of this new pathway to the future.

Global peace will not be just peace on earth but will require ways to insure peace in the skies as well. The new space economy will need to be built not just on technology; it will also need to be founded on strategic space defense systems as well as international space agreements and a new "star map" to the natural resources on the land, the seas, and also the cosmic seas above. To achieve this new abundance, all the people of Spaceship Earth will need to create a new bond of cooperation, with legal and regulatory mechanisms, backed by strategic space systems to insure that this will ultimately be made to work.

This will likely all start by reviewing the global pacts and agreements that we have used with regard to cooperation and sharing in the global commons represented by the oceans and Antarctica and even outer space that surrounds Earth above commercial air space but below the vacuum of outer space—an area that is sometimes called subspace or protospace. If we can build on historic patterns of cooperation in these global commons, then there is potential that we can build a global governance system that can allow the era of astral abundance to become reality—sooner rather than later.

No one should be naïve enough, though, to overlook the fact that our future in space involves a three way tug-of-war between: (1) the new businesses hoping to realize the riches of outer space via space commerce; (2) the push for new international space agreements—i.e., new "rules of the outer skies" and cooperative space standards and practices—that can allow a fair and equitable set of practices for the "cosmic commons" and (3) the strategic and even military space systems that will "police" the new space economy as it grows and matures further and further away from Planet Earth.

Astral Abundance

Unless we turn to the commercial opportunities of New Space and breakthrough new technologies here on Earth, we could indeed be in deep trouble. This will be ever clearer as populations continue to rise and resources shrink (Fig. 1.3). Dr. Thomas Malthus, the economic prophet of the eighteenth century who predicted we would eventually run out of food and vital resources, will be proven right even though he was perhaps three centuries premature. Some very capable people have gathered data from all over the world to put together the following chart on so-called "non-renewable resources."

This chart was compiled in 2000, when the global population was under seven billion and commodities were more abundant. Think what this chart will look like as time goes by, as world populations grow and as global standards of living rise and everyone in the world wants a computer and a smart phone.

Global Natural Non-Renewable Resource Scarcity			
Extremely Scarce	*Very Scarce*	*Moderately Scarce*	*Not Currently Scarce*
Bromine	Aluminum	Antimony	Arsenic
Gold	Bauxite	Beryllium	Barite
Mercury	Cadmium	Bismuth	Boron
Tantalum	Chromium	Cobalt	Garnet
Tellurium	Copper	Gallium	Lithium
Thallium	Fluorspar	Germanium	Niobium
	Magnesium Compounds	Graphite	
	Molybdenum	Gypsum	
	Nickel	Indium	
	Rhenium	Iron	
	Selenium	Lead	
	Strontium	Lime	
	Sulfur	Manganese	
	Tungsten	Salt	
		Silicon	
		Silver	
		Tin	
		Vanadium	
		Zinc	
		Zirconium	

Fig. 1.3 Global inventory of scarce global resources

However, astral abundance is about more than just resources and energy. It is about new and different types of jobs. In short, it is about new hope, new horizons, a new understanding of sustainability, new frontiers for humankind and a new understanding of the shared global bounty of the riches of outer space .

The new space and advanced technology economy will bring the opportunity for new and different types of jobs, better global education and health care, a cleaner and healthier environment, and what Pope Francis would call the "bright light of hope for a better and more peaceful world."

The New Space Entrepreneurs and Commercial Space Enterprises

The new world of astral abundance can bring us a wide range of totally new jobs for the millennials and even a redefinition of wealth. Today there are only a small number of people now working at Planetary Resources, Deep Space Industries or Shackleton Energy. These young and enterprising people are focused on ways to mine important resources available in outer space that might be available from near-Earth asteroids.

Some enterprising workers in the New Space industry are busy designing solar power satellites that can bring us clean and abundant energy from outer space. Others at SpaceX, Bigelow Aerospace, Virgin Galactic, Boeing, Sierra Nevada, Blue Origin, British Aerospace, XCOR Aerospace, Orbital Sciences ATK, Kelly Space and Technology, Swiss Spaceplane Systems, Reaction Engines Ltd, Stratolauncher, etc., are seriously working to engineer and build new spaceplanes and commercial launch systems These are just some of the new cadre of New Space enterprises working away to create the new commercial space industries.

Meanwhile enterprises such as Lockheed Martin Skunkworks have shown us how we can unlock the energy of the Sun right here on Earth. Others such as the Blue Brain project, headed by Professor Henry Markram in Switzerland, are busily seeking to create artificial intelligence with the equivalence of the human brain. The synergy of the all these enterprises will be critical to the ultimate achievement of this breakthrough economy. This breakthrough to a new type of world has been anticipated for some time. However, it would not be possible without first achieving new international agreement and standards of behavior based on true corporate responsibility.

Breakthrough: Transcendence? The Singularity or Astral Abundance?

The terms "the Singularity" and "Abundance" are used interchangeably throughout this book to refer to breakthrough technologies and the rise of super intelligence that are presumed to accelerate global innovation and the means to cope with problems of all types, from clean energy to climate change to overpopulation. It was Ray Kurzweil, the Artificial Intelligence (AI) guru, who popularized the term Singularity. Dr. Peter Diamandis, who in cooperation with others founded the International Space University and then went on to found the Singularity University and also breathed life into the wonderful X-Prize initiative, simply calls it "abundance." And before him R. Buckminster Fuller called it "transcendence."

No matter what you call it, the idea is to go ahead and think outside the box. Indeed the trick is to think outside the limits of the 6 sextillion-ton spaceship we call Planet Earth. Fuller, Kurzweil, Diamandis and other space enthusiasts, including the authors, are trying to convince our economic and political leaders that the trick is to think outside constraints of the current world economic systems and the resources we have trapped within the orb we call Earth.

The key to such a breakthrough may rely not on the graybeards but on the youth of the world. Many of those under 30 in age are more than a little miffed at the older generations and the mess that they have created here on Earth. They look askance at pollution, greenhouse gases, exhaustion of resources, religious strife, unlivable mega cities and excessive and runaway population growth. The graduates of the 2015 Space Studies Program of the International Space University have started a series of meetings around the world called "Disrupt Space" in localities such as Shanghai, China, and Bremen, Germany. Their idea is to challenge a group of international entrepreneurs to work with corporate representatives, government officials and investors to combine efforts to solve today's problems using space. The full title of this new space global initiative is: "Disrupt Space Summit: Turning the Solar System into Our Backyard. Let's Play!"

The problem with such initiatives is not innovative thought, not youthful zeal, not lack of commitment to real change. The biggest problem could turn out to be human arrogance or a warped value system that puts expansive growth above societal survival. We need real understanding of "sustainability and long term ecological reform" before we humans venture seriously into space with the mistaken idea that a larger playground ensures a better tomorrow. We must embrace the long-term sustainability of both Earth and space as we venture forth toward a better tomorrow.

We must embrace the potential of the Solar System, of solar energy, of nuclear fusion, of artificial intelligence, and examine what is meant by the breakthrough economics of transcendence, the Singularity or astral Abundance. If our political, economic and business leaders are able to think creatively and embrace the potential that the space billionaires and New Space entrepreneurs can unlock, then we can solve the problems of a world populated by ten billion people and over-urbanization.

Today there are just a few hundred highly entrepreneurial companies that understand the potential of astral abundance. These visionaries are hell-bent on engaging in space mining, solar power satellites, nuclear fusion, artificial intelligence and robotics, and transcendent technologies that can free us from the non-sustainable practices of past industrial practices that exploited resources and the ecology rather than the potential of human intelligence and astral abundance. If one goes to the website of Planetary Resources, Inc., you are met with the following evocative message: "We dare you to change the course of humanity with us." Peter D. certainly never thinks small. If you should be able to get into the ever more selective Singularity University (founded by Peter Diamandis, Ray Kurzweil, and Peter Worden, until recently head of NASA Ames) the challenge you are given is to come up with an idea or project or invention that will have a positive impact on a billion people within a decade's time.

Those that are starting up new companies to engage in space mining today are largely focused on rapid payoffs to fund their larger and longer term ambitions. They hope to realize a significant economic return from exploiting rare earth minerals, platinum-rich asteroids and new energy sources such as helium-3.

The enterprising Planetary Resources Ltd. was able to raise $1 million from crowd sourcing for their Arkyd space telescope that will seek out promising near-Earth asteroids that could be captured and brought back to Earth or lunar orbit. Not only is this a huge technical and engineering challenge, but there are a range of legal and regulatory questions as to whether such action is consistent with the Outer Space Treaty of 1967 to which more than 100 countries are members. While such issues are being considered in the United Nations, Planetary Resources Ltd. forges ahead to cope with the problems they feel technically competent to address (Fig. 1.4).

Fig. 1.4 Arkyd's small space telescope is used to locate promising near-earth asteroids (Image courtesy of Planetary Resources Ltd.)

Keys to the Era of Astral Abundance

But what the commercial space explorers and entrepreneurs must ultimately do is more than answer the questions of why, when, where and how space mining is to be done. In the end they must help unlock the potential of a new astral economy based on the full riches of the Solar System. These riches can and will include clean, limitless, and low-cost energy, a stabilized population that comes from a high standard of living, a smart robotic work force, and universal education and health care. This may sound like a world for dreamers. But it is the space entrepreneurs of the astral abundance economy that will turn this dream into reality by sustained and energetic enterprise.

Some of the entrepreneurs are planning to build solar power stations in the sky that can beam back clean energy to Earth on a 24/7 basis to meet our future power needs in the latter half of the twenty-first century. Other enterprises involve colonies on the Moon, private space hotels and habitats in the skies. Some, including the current authors, have envisioned other exotic ventures such as systems that can protect us from solar storms while also economically providing Earth with abundant clean power.

Meanwhile, back on Earth, scientists can also develop fusion-based power plants that can not only light up cities but also allow systematic desalination and mining of the oceans. Yet others can concentrate on developing smart robotic devices, with the IQ of 100 or more, that can work in industrial production and the service industry to free up humans for artistic, design conceptions, scientific and biological research, and philosophical studies. If such capabilities are widely shared within the global and astral commons, the need for warfare and aggression can be minimized. Indeed the number one question for the age of astral abundance is whether warfare can be phased out if abundance can be made universal.

Clearly the world today is not prepared for astral abundance. It is not clear what would be the economic, social, cultural, political and religious motivator in a world where everyone could have all they need. Today's global economic system is still based on economic need and competition between haves and have-nots. Today's economic systems and incentives would fail if everyone could have all the food, housing, transportation, education, and health care they wanted while smart robotic devices were available to everyone to carry out routine farming, industrial production and ordinary service functions. Communism failed due to a lack of incentives to work and to take pride in their work. Humans are conditioned under capitalism, and even socialism, to work and strive for recognition, and to achieve. Clearly the challenge in a world of astral abundance and artificially intelligent robotic workers, is finding ways to motivate people to strive and be productive outside the old patterns of work, wealth accumulation, status, civil, political, economic, cultural and religious strife and old patterns of unrest and warfare.

Developing Commercial Space Transport Is Only Phase One

The key to all this is new commercial transportation systems offering lower and lower costs for access to space. Elon Musk at SpaceX, Sir Richard Branson at Virgin Galactic and Paul Allen, providing the backing to build the first spaceplane and the giant Stratolauncher, are just three examples of space billionaires creating the new commercial systems that can go cheaply into space.

The world community is slowly beginning to recognize that space is not some luxury to spend money on to explore, but it is truly the next frontier—a frontier whose settlement, use and exploitation in sustainable ways may be our very pathway to ultimate survival as a species.

The essential point of this book is that New Space is the next true frontier—not only for the space billionaires but for the entire human race. The new "gold rush in the skies" is not just a slogan but an essential pathway to the future. It is our ultimate hope for survival of the human race. This is why billions of dollars—that's billions with a "b"—are now being invested to create a new space economy. And new rules and regulations are being put into place to make this new gold rush a financially viable and competitive enterprise. This is not speculation about the future; it is part of a wide range of New Space activities that are starting to happen NOW!!!

New Jobs, Resources, Energy and More—The Door to the Long-Term Cosmic Future

We are talking about new jobs, new resources, new energy systems, and even new mechanisms to cope with the planetary needs of a planet whose resources are today being over consumed. There are scores of metals, rare earth minerals, and other natural resources that we are currently exhausting. Despite this overconsumption, our global human population continues to grow apace.

If we can succeed in conquering space in new and creative ways, then there is a chance that humanity may be able to sustain itself not only for hundreds or thousands of years into the future but perhaps for eons to come. We might even be able to outlast the Sun!

However, to do so we must find ways to utilize the riches of space in better ways than we have managed to do here on Earth to date. We will also perhaps need to generate potable water and protect our planet from solar storms and asteroids, and strengthen a shifting magnetosphere to shield us from cosmic dangers.

In short, our very future may depend on going into space in innovative, intelligent, and renewable ways. Our survival within the next 100 years may depend on the creation of a new space-based economy that involves the mining of the Moon and asteroids, creating solar storm shields in space, and perhaps even embarking on other grand enterprises such as the colonization of the Solar System. These new space enterprises will likely involve the bringing of minerals and waters trapped in asteroids and other celestial bodies back to Earth as well as building a new global economy, largely based on renewable resources and entirely new concepts in urban planning and ecological engineering. It will also involve harnessing corporate enterprise and innovative thinking among the New Space entrepreneurs to allow all this innovation to occur rapidly and in a fair and equitable way so that the entire human race can feel they are a part of this post-industrial, New Space economy.

The Key Elements of the Post-Industrial New Space Economy

There are lots of things that we can expect to happen in the New Space economy that, if not today then within the next two decades. Important new capabilities and industries will join today's multi-billion dollar space industries that today include satellite communications, remote sensing, and space navigation. These coming New Space industries will include:

- **Space Transport and Low-Cost Access to Space:** Commercial launch systems and hypersonic transport across the oceans. These systems of the 2020s may in time give way to even lower cost and more reliable ways to get into space. These might include tethers or so-called space elevators, or mag-lev systems. Many of the key new technologies can be used here on Earth, the Moon or other planetary bodies.
- **Space Mining:** This would involve getting, at reasonable cost, a new source of minerals, metals and volatiles (i.e., water and gas) that we are either using up here on Earth or are expensive to launch into space. We need to find ways to recycle the materials we are currently using up on Earth and consume natural resources in more sustainable ways. In short, as we engage in space mining we need to explore in parallel other capabilities such as ocean mining and improved recycling capabilities.
- **Clean Energy from Space:** There are various ways to get clean and significant energy from space. This could be helium-3 or other suitable isotopes from the Moon or other sources such as asteroids plus solar power satellites. Of course many think beaming solar power from space back to Earth can also be a viable space business.
- **New Infrastructure and Habitats in Space:** This could be facilities to protect us from asteroids or solar storms, habitats or facilities for space robotic production, and other dwellings in space to facilitate everything from space tourism, space manufacture, or even space colonies.
- **On-Orbit Repair, Satellite Upgrades and Active Debris Removal:** These capabilities can support commercial communications, remote sensing, navigation and timing, and other space services. Debris removal may become much more important as debris in low Earth and polar orbits continue to build up. These capabilities could also have strategic implications, since such capabilities might be considered to be weapons systems.
- **Strategic Space Systems That Can Enforce New Cosmic Space Enterprise Agreements:** The story of human history and problems of corporate malfeasance suggest that there must be enforcement procedures and sanctions

to safeguard the New Space agreements that enable the new space economy and astral abundance to move ahead.
- **Space Products and Industries We Have Yet to Imagine:** When batteries, electricity, lasers, even Velcro were first invented the applications were far from clear. This will also prove true for New Space technologies as well. The new technologies and systems created for New Space services on the Moon or asteroids will have many spin-off applications. Prime candidates for new uses include advanced robotics, remote power systems, space tracking and guidance systems, and remote telecommunications and information technology systems.

There are many technical, engineering, business and economic issues that will need to be solved before any or all of these new space industries are fully realized, but in order for commercial space to move ahead, there will need to be a new space agenda agreed by the many countries that are members of the United Nations and the U. N. Committee on the Peaceful Uses of Outer Space (COPUOS). Today many of the space treaties and conventions agreed in the 1960s and 1970s are out of date. New agreements, rules of conduct in space, transparency and confidence building arrangements or other codes of conduct are needed to address the duties and responsibilities of commercial entities, international consortia and other actors in space.

Clearly some new arrangements or new forms of space regulation may be needed with regard to the future use of the Moon, near-Earth objects and other planetary bodies. Some regulatory certainty can help create the New Space economy if those regulations create a positive framework for progress. Such reforms in international regulation related to space would presumably allow the various inhabitants of our planet to share in the benefits of the New Space economy but would avoid putting the kibosh on innovation and entrepreneurial zeal. The key is to spur new technical and economic innovation but in a way that is equitable to all inhabitants of the planet. A delicate balance here must be achieved, and one can only hope that the space community is allowed to work this out without becoming enmeshed in the broader macro politics of our times.

The issues related to space enterprise and regulatory policies for activities that will soon be conducted well above Earth's gravity are now beginning to receive more and more attention. The Commercial Space Transportation Act of 2015 in the U. S. Congress represents current thinking about mining for resources in space. This is just one legislative measure, but it is far from the only one being considered. Echoing national legislation in Europe and Asia is almost certain to follow. The key question is whether this can lead to some sort of global agreement or possibly some "soft law" agreement such as a new "rules of the road" for commercial space activities [3].

Today it is far from clear as to how space mining will be carried out in outer space. No one really knows how these activities will be regulated and how benefits might be shared in the space bonanzas yet to be realized. What is clear is that it will take years—and possibly decades—to work out the answer to everyone satisfaction.

Conclusions

What should be realized is that if we can develop smart enough technology, the natural resources to be harvested from the great beyond are vast and almost unimaginable. There are likely over one million asteroids that are 30 m or more in diameter that can be accurately classified as near-Earth asteroids. This means that for each country in the world there are thousands of asteroids out there for future possible harvesting of resources.

The question therefore is not, is there enough for everyone. It is rather whether we can we find an effective and equitable economic model that works. We need a "business model" that does three things. It must allow us to capture and utilize the resources so that everyone on Earth can benefit from these astral assets. It must allow the space pioneers that make it happen to be able to profit from this enterprise and recoup their huge capital investments. And it must be done in an environmentally and ecologically safe and sound fashion that does not pollute space, create adverse effects such as climate change, or endanger life. Indeed if carried out in the right way it could even help to ameliorate climate change.

The above principles are clear. The achievement of such goals, however, is indeed very hard to realize. The space zealots say: "Just do it!" The skeptics and non-spacefaring nations quite logically say: "Not so fast. We need iron-clad guarantees and well-conceived plans."

The real thing to focus on is the wealth of opportunity that is potentially available to humankind. We need a completely new mindset. There are indeed vast gold mines in the skies. We need to look at space as a new frontier that opens almost unlimited opportunity.

Just what is out there? Well we know there is unlimited energy that can come from solar power that is available 24 h a day and never shielded by clouds. We know there are key isotopes ideal for fueling fusion engines.

We know that there is water that can be broken down to hydrogen and oxygen to fuel spaceships. We know that there are rare metals such as platinum and other valuable ores that are being depleted here on Earth. We think there are enough resources reachable within a fraction of one astronomical unit, sufficient to replenish many of our rarest natural resources.

We know that the technologies we need to develop in order to acquire and use the available energy and other resources can also bring enormous spinoff benefits to everyone here on Earth as well. In short, the new technologies we develop are critical to progress. These new breakthroughs in space transportation, hypersonic travel, robotic mining, remote refining and production, solar power satellite systems and so on, can support our outreach to acquire cosmic resources. Just as the wheel, the lever and the pulley allowed humankind to evolve thousands of years ago, the technology of the New Space economy will propel us into the twenty-second century.

The most effective and equitable economic models involve crucial balances. They need to include means to recover capital investments and research initiatives. They will involve reasonable sharing without unreasonable requirements that would block investment in new levels of new space commerce. This is easy to conceive in theory, but will be very hard to negotiate and agree to in practice. Space agreements related to sharing of clean energy, mining and selling of resources, coping with climate change, non-military uses of space, and all forms of space commerce are critical aspects of the new age of space commerce and shared abundance that are quickly becoming critical to the future world of cooperation, exploration and resource sharing. The provisions of the new study called the Global Space Governance are seen as a potential blueprint to start this new round of cooperative space agreements. The hardest part, ultimately, will not be to envision these new "rules of the skies" but to devise a system of enforcement that is economically viable, effective and broadly accepted by the entire international space community.

References

1. Disrupt Space Conference, April, 2016. hi@disruptspace.io. Last accessed May 16, 2016.
2. "If there were a day without satellites," 2016. https://www.youtube.com/watch?v=5sgM7YC8Zv4.
3. Senate Passes Compromise Commercial Space Bill – UPDATE, November 11, 2015. http://www.spacepolicyonline.com/news/senate-passes-compromise-commercial-space-bill.

2

A Space Cornucopia of Jobs, Resources and More

Introduction

There are those that quickly accept defeat when confronted with what appears to be an insurmountable challenge. Others, when faced with significant change in circumstance, naturally tend to fight back in order to preserve what they have achieved. Then, there are those unique individuals that transcend time. They create a new future, before the general public realizes that things have changed forever.

These innovators are the ones that create disruptive technologies, such as Uber, Lyft, Amazon, the Internet and satellite and cable-based electronic entertainment. The space entrepreneurs are like time travelers with their feet in the twenty-first century.

Space age innovators are intent on creating the New Space economy, and they are half Thomas Edison and half John D. Rockefeller. These "transcenders" of today's technology and business practice are remarkably different in personality and background as one can imagine. Despite these differences, they are alike in boldly viewing the world—not as it is but as it can be.

They pursue their various New Space enterprises—not simply to make money or to create a successful new business. No, they want to change the world and recast humans with a new image. They see Homo sapiens not as farmers and manufacturers and service providers in a closed world, but rather as masters of time and space. They truly believe that a New Space cornucopia will not only release new riches but fulfill human destiny. They are thus embarked on their various missions in order to allow the tribe of Homo sapiens to survive. They want to spread the seed of humans across the Solar System and eventually beyond. They earnestly see the "mission" as boldly going where no humans have gone before. The short mission statement or goal is to ensure that the best days of humanity are in the future and not in the past.

Robert Bigelow, Peter Diamandis, Jeff Bezos, Elon Musk, Paul Allen, Eric Anderson, Ray Kurzweil, and Sir Richard Branson are among those that are busy seeking to create a new future of space abundance. They are 100 % sure that new technology, human innovation and the unlimited potential of outer space will allow us to reinvent the global economy.

Despite their optimism, it is possible that even these visionaries do not fully grasp the broad scope of change that the New Space economy will bring to the world. Indeed the New Space cornucopia has enormous range and competitive advantage.

Peter Diamandis envisions low-cost space travel and mining the asteroids. Robert Bigelow envisions four-star hotels in space. Paul Allen envisions the world's largest jet that will serve as a launching station for rockets of the future. Eric Anderson envisions private space launches that can go to the Moon and back. Ray Kurzweil envisions solar power systems, including solar power satellites that make global energy clean, plentiful and pervasive. Elon Musk sees a million people living on a space colony on Mars. Sir Richard Branson named his space transportation company Virgin Galactic because he truly thinks not only outside the box but outside the "circle" that is Earth.

If you want to image the full scope of change that all this could bring, let us engage in some time travel. Try to imagine our world today without the Internet, personal computers, Instagram, Facebook, smart phones and spandex. Then recall that these services or products were essentially unknown just 30 years ago. Or let's go back a century and imagine a world where airplanes, automobiles, television, electric lighting and broadband communications were either unknown or were available to only a few. For millions of years life on Earth was relative static and changed very little. But in the age of rapid human innovation and the Internet it seems as if it changes every few minutes.

The truth is that we live in an age of super time compression, and innovation is not only speeding up but the rate of acceleration is actually increasing. This phenomenon physicists call "jerk." The bottom line is that in the age of "New Space" and electronic innovation, that change is coming faster than anyone thinks today. A few years back, the author wrote a book called *e-Sphere: The Rise of the Worldwide Mind*. This book is about e-Space, a new age of human expansion into a cosmic world and New Space economy that transcends the world as we know it today. "E-Space" will become a new reality within the next half century. It seems that only major disruptive innovations variously described as "the singularity," (Kurzweil) "abundance," (Diamandis) etc., can alter humanity's future evolutionary course, or we are indeed headed into some major global troubles. These global problems include such challenges as climate change, overpopulation, intensive urban crowding, lack of potable water and a host of other problems related to over consumption of resources.

Indeed these current large-scale global problems present what might be called a cosmic ethical dilemma—whether we need to show that we can cope successfully with climate change and perhaps demonstrate that we can devise a successful artificial "biosphere" here on Earth before we try terraforming planets elsewhere, or whether we should just move ahead. And by devising a "successful biosphere" we mean show that we are smarter than when we failed to keep "Biosphere II" viable for "biospherians" to live inside after only a few months of experimentation. Until we can show that we truly can cope with climate change and create an artificial habitat that can sustain itself over the longer term without killing off "biospherian" life forms inside, it does seem presumptuous to say we know how to create viable colonies on Mars or the Moon or elsewhere off the planet.

Of course, this is not only the challenge of the twenty-first century. There are also the challenges of finding ways to create sustainable global economies and global employment. This becomes a huge problem when smart machines are increasingly supplanting more and more jobs—especially in the new service economies. Some would argue that technology is now solving old problems but creating new ones at a record rate.

However, one can be optimistic that we will eventually become technologically and environmentally savvier. The grand challenge is to achieve long-term sustainability and discover what might be called environmental prosperity. If this can be achieved it will be the ideal time to look off planet in a truly permanent fashion. If we can develop the intellectual maturity to create a sustainable world then we might be ready to create off world civilizations as well. In short, the key step toward off-world colonies is to develop a new kind of technology that not only produces gizmos, gadgets and economic throughput, but true long-term societal solutions.

New Space and Creating Jobs

One of the keys to this new world rich in sustainable technologies is to find how to create productive employment for humans in an automation-rich world.

One of the big political debates in the United States today involves the issue of minimum wages. Democrats, with their low to middle income constituencies, are arguing that a raise in the minimum wage not only provides a "livable" standard of living but that such a policy would also stimulate the consumer economy. This seems fairly reasonable in that people who live on a minimum wage must spend virtually all of their disposable income. Republican advocates on the other side of the discussion, with more of an eye to their business-oriented constituencies, say that a large rise in the cost of living would be inflationary. They would argue that such a policy will only give rise to increased automation. They quite reasonably argue that, since "basic human labor" is becoming too expensive in comparison to machines, raising the minimum wages will lead to a death spiral of basic labor jobs.

The indisputable truth is that automation, artificial intelligence and expert systems are now able to replace more and more jobs. These automatons, smart robots and expert systems are not only replacing routine manufacturing, farming and mining jobs but an increasing array of more skilled service jobs as well. Just in the last few years, robots and AI computer programs have surged in numbers and sophistication. These increasingly capable machines and heuristic programs that more and more simulate the skills of "thinking humans" are moving into the service workplace. Such "smart systems" are replacing tax accountants, real estate appraisers, auto workers, pharmacists, and those that carry out inventory control in factories and warehouses.

The unanswered question of our time is, what do people do when machines take over the majority of all jobs on the planet? The minimum wage debate misses the key point that all sorts of jobs are disappearing and that we need to reimagine employment in a super automated world [1].

This is a profound issue with over a third of the world's adult population either unemployed or underemployed, and the trend line headed in the wrong direction. More and more people and fewer and fewer jobs is not a formula for success. The road to sustainability must lead in a different direction. That much is a no brainer.

Does the creation of new space colonies, space enterprises and new off-world economies and settlements become a part of the answer? Only time will tell. But this much seems clear. Time and human technological progress is a one-way gate. The arrow of time points in only one direction. We are trapped in the future we invent.

Certainly, the creation of a New Space economy and prospect of off-world activities represent the only major truly new human enterprise of our time. This is just one of the reasons that new entrepreneurial enterprise related to outer space services and manufacturing is currently quite popular in the U. S.

Congress. "New Space," being a totally new source of jobs plus its powerful stimulus to new technology and totally new types of endeavors, led to the Space Act of 2015, with its positive incentives and stimuli such as governmental liability coverage, rights to individuals to own resources mined in space and so on.

The trend to support new space endeavors is apparent around the world. Over two dozen countries have implemented or are planning spaceports. There are over a dozen spaceports now approved by the FAA in the United States. And there are a number more in the pipeline.

Spaceplane development is active in a growing number of countries around the world. Part of the interest and support for space activities is the appeal of high technology research and development and being on the cusp of new discovery and invention. The most important reason for support, however, is the prospect of new jobs.

In so-called developed economies, less that 3 % of jobs are now in the farming and mining sectors. Jobs in manufacturing continue to drop. About 12 % of jobs in developed countries are in manufacturing, and of these "manufacturing" jobs an increasing number are in areas such as sales, promotion, management, engineering and design rather than actual manufacturing. This means that 85 % of the jobs in developed economies are in services. Yet as just noted these service jobs are increasingly being automated or turned over to devices or robots that are artificially intelligent.

Ray Kurzweil, the artificial intelligence guru that invented "Siri," who so sweetly and competently responds to inquiries on smart phones, believes that the "singularity" is coming within the next few years. The term "singularity" was first used by John von Neumann in 1958. It was then amplified by Vernon Verre of Hungary and even more recently given a more focused meaning by Kurzweil, especially in his book *The Singularity Is Near*, published in 2005.

Kurzweil predicted high speed processors, memory storage and artificially intelligent algorithms that would not only duplicate human reasoning, memory and processing capabilities but would be commercially available at $1000 per unit by 2029. He anticipated that by 2040 there would be remarkable new capabilities in machine intelligence. These new AI capabilities in his world view would have an exponentially profound impact on human civilization. This he characterized as the "singularity." New Space technology and systems working in conjunction with breakthrough "singularity," such as AI and computer technology, could, in his view, transform and disrupt every aspect of the last half of the twenty-first century.

At the Comsat Corporation in Washington, D. C., in 1969 when Kurzweil was asked what is the most important and transformative technology of our time, without hesitation he came back with the answer "artificial intelligence." The audience was expecting space or communications or perhaps satellite communications. But Arthur C. Clarke, who gave us "HAL" in the movie *2001: The Space Odyssey,* was resolute in believing that thinking machines was the technology that would change everything.

If Kurzweil's projections are in any way accurate the future of human history, economic systems, and employment are headed for profound and perhaps explosive change. In the new post-2040 world, the impact of smart robots, reasoning and highly trainable machines will likely change employment, wealth, economic structures, and patterns of life in a significant way. What this means with respect to the New Space economy is incredibly difficult to project. But certainly it can redefine the human future. The range of the dynamics can be quite wide. The concept of Elon Musk, that there will be a million-person community living on Mars, able to sustain the full range of industrial, educational, and intellectual activities comparable to that experienced on Earth, is one bold vision of the future. The one thing that seems clear is that the nature of human activity in the twenty second century will have changed forever. Part of that change will likely be some form of off-world activity.

There are many ways to project the future and note trend lines. One interesting statistic is that when NASA called for candidates to become astronauts in 2015 there were 6300 people that answered the call and four men and four women were selected. In 2016 there were 18,300 candidates that answered the call. At least young people seem to think there is an important future in outer space.

All that people need in order to subsist and procreate at a basic subsistence level is food and water, basic tools and protective devices, housing, and energy. To improve their lives they need health care, education, clothing, transportation, and perhaps art and culture. Only a small percentage of the work force is now needed to supply all these needs.

Currently automation is responsible for the bulk of the "work" needed for subsistence in so-called developed economies, and this also became true for manufacturing in the 1950s and 1960s. What might be called "super automation" that comes from machines or robotics with expert systems or artificial intelligence software seems capable of replacing most service and skill level jobs. According to Ray Kurzweil this will all happen by 2040 [2].

If we look at today's current reality this seems entirely feasible. IBM's Watson, using so-called Unstructured Information Management Architecture (UIMA), is now able to beat the world's Jeopardy champions, and currently the latest focus at IBM is to make Watson the repository of all known medical knowledge. According to IBM researchers, medical information (and presumably knowledge) will be doubling every 73 days by 2020. In such an environment it indeed seems likely that only a machine that can cope with processing petabytes per minute could possibly contend with this plethora of information [3].

Apple's SIRI is able to answer most questions that people would have in their modern everyday life. In the world of the "singularity" in which low cost and amazingly "smart" digital processors cost under $1000, our world, employment and the nature of work will change in a way that we are totally unprepared to understand. Without the benefit of a new challenge, human civilization stands at risk of stagnation or worse. The image of the future provided by Kurt Vonnegut, Jr., in *Player Piano* is a very frightening one. In this sci-fi book the members of the vast proletariat are simply consumers, and only a handful of scientists and engineers have jobs that consist mainly of keeping the machines running. It is a frightening dystopian forecast for the future.

This view of the future that Western technology seems to be delivering to the world is presumably part of the prospect of the future that extreme Jihadists are fighting to undo. There is comfort to some in setting the clock of technological progress back a few millennia. Al Qaeda and other technological nihilist groups see comfort in a human return to a world of the brutish caveman.

What will the world of the future be like, in fact? If by 2200 there were viable human habitats in Earth orbit, true colonies on the Moon and Mars, plus mining operations on large-scale near Earth objects (NEOs), our future could be dramatically different. This future would be dramatically changed not only in terms of work but also in terms of broadband communications, clean energy systems, and our knowledge of the universe and our creation of sustainable off-world living environments. In short, the new gold rush is about more than space technology. This new future in space will alter most aspects of human life as we know it today.

A New Source of Natural Resources

Nearly a decade ago the space agencies of the world agreed to work together to combine all of their collective data represented by satellite Earth imaging. In this manner they created a very high resolution integrated image of the world. This amazingly comprehensive image of all of the world's forests, swamps, waterways, deserts, mountains, oceans, glaciers and icecaps represented many terabytes of data.

What was even more important was that we were able to create comparative global images over time that mapped exactly how the surface terrain of the entire world has changed over 360° of latitude and longitude.

These high-definition images showed a world that is becoming increasingly urbanized with expanding human settlement. They showed how the deserts of the world are increasing and areas of vegetation and forests are shrinking. The collective results of these satellite-based images were to show a world at increasing risk. In short there is more to worry about than human-driven climate change, but rather to understand that human development and relentless growth gives rise to a number of problems that are interlinked together.

Today there is more and more remote sensing data. The images are now increasingly precise and insightful, with the addition of hyper spectral imaging. Today there is not only data from space and governmental agencies that collect meteorological and oceanic data but also from Google Earth and private data collectors such as GeoEye. For those who wish to stare reality in the eye there is a great wealth of imaging data mounting into the petabytes (i.e., thousands of trillions of data points).

The unpleasant truth of how the world is changing is revealed by satellite and UAV imaging. We are no longer seeing changes from the broad vistas of continents, but relentless change acre by acre, meter by meter in amazing detail. And of all these elements of change, one transformation is by far the most profound and disturbing. This change is the amount of water available to humans and to animal and plant life. Here we are speaking of precious water that is not saturated with salt, the non-seawater. This untainted water that sustains life across our planet is shrinking. And along with this diminishing water supply comes less vegetation and fewer forests and increasing deserts.

The latest assessment of the scope of change over the last three decades suggests that thousands of square kilometers of once verdant areas have changed to deserts and barren fields. Our deserts are growing and our farmlands are shrinking. Our remote sensing satellites such as those operated by the commercial operator GeoEye lets us obtain images that are on the order 35 cm^2 per pixel, or one dot per sq. foot of visualization. This is sufficient to discriminate between cars and SUVs and to determine what types of crops are growing in fields. Figure 2.1 below shows us the Saharan Desert in detail as it encroaches on land that was once farmable.

There is more to these broad global changes than just desertification and shrinking water supplies. There is also the loss of the rain forests, swamps and everglade areas, and the melting of icecap areas. The melting ice in the tundra of Siberia serves to dissolve peat fields there that ultimately will release billions

Fig. 2.1 The Saharan Desert in the Sahel area of Africa keeps expanding (Image courtesy of GeoEye. http://www.satimagingcorp.com/gallery/geo-eye-1/geoeye-1-sahara-desert/.)

of metric tons of methane into the atmosphere. The consequent release of methane from these peat fields is at least as significant as the release of carbon dioxide from coal-fired power plants. The implications from all these changes are worrisome at best and potentially disastrous at worst.

In terms of implications for human life, the shrinking water supply is currently the most concerning, even though the other changes are also highly destructive as well. No place on Earth today is turning greener. Across the planet land is turning browner. Around the world there are more people and less water. Figure 2.2 dramatically visualizes what an overpopulated world will come to look like, but the problem that will define our future will not

Fig. 2.2 Crowded beaches are just one signal of an overcrowded world (Image courtesy of Essential Environmentalists.)

be the lack of a place at the beach but insufficient water to drink as global populations grow from about 7.3 billion today to somewhere between 10 billion and 12 billion by 2100.

An even more worrisome image is that of villagers in Ethiopia crowded around a water hole that is growing dry. The image below in rural Ethiopia is being repeated across the Sahel region of Africa, where water holes and depleted aquifers below the surface that once carried a huge supply of fresh water is drying up and ending the viability of village after village. One estimate is that as many as 20 million people have felt the harsh hand of climate change and the drying up of water supplies that have forced the relocation of villages in Africa and indeed in Asia and South America as well (Fig. 2.3).

The water that is so abundant in our oceans provides us with a false impression of the supply of fresh water that is essential to life and the cultivation of crops in farms around the world. The amount of fresh water that is accessible is a tiny fraction of the water in the oceans. And as the icecaps melt and turn fresh water into saltwater, our supply is shrinking. The following graphic prepared by the Sierra Club shows the amount of water currently available in rather stark terms. The lack of water is one of the major growth boundaries that humanity faces today (see Fig. 2.4).

Space mining of the skies and significant amounts of desalinization can change all this, but to understand the logic of space mining we need to

Fig. 2.3 An Ethiopian water hole that is muddy and drying up in the Sahel region of Africa (Image courtesy of Engage Now Africa.)

understand much better the makeup of the Solar System and what resources are out there. Most people know enough about the Solar System to understand that there are the Sun, the planets, their moons and the Asteroid Belt. The general public does not necessarily realize that there is between 500,000 to 1 million so-called near Earth objects that come within striking distance of our world over time that are big enough to be city killers—or over 30 m (some 100 feet) in diameter. These potentially hazardous asteroids that come within 0.05 astronomical units, or about 7.5 million km (4.7 million miles) are now the subject of an intense hide and seek operation using ground observatories around the world and infrared space telescopes.

Fortunately most of the massive dangerous asteroids that are over a kilometer in size have been found, but there are still a huge number of the smaller "city killer space rocks" still to be located and their orbits charted. The point is that there are many, many more space rocks out there than most people realize, and that these rocks contain valuable minerals and metals (such as platinum) and what scientists call volatiles and we call water. Out in space we could use this water for rocket fuel (i.e., breaking water down into liquid hydrogen

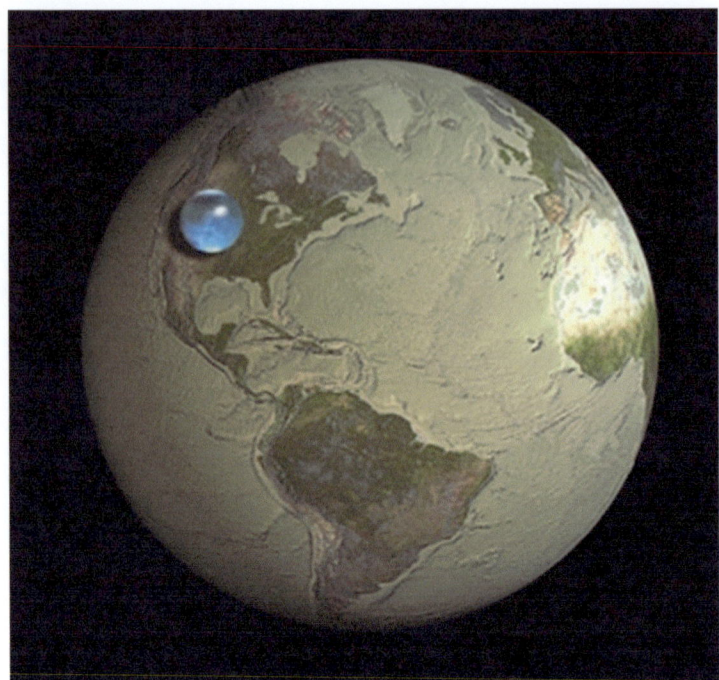

Fig. 2.4 The small amount of pure water available to humanity today (Image courtesy of the Sierra Club.)

and liquid oxygen), to sustain space colonies and for other key purposes. The current theory is that the water that constitutes our oceans came from asteroids or comets and that there is far, far more water in outer space than in the Atlantic, Pacific and Indian Oceans—tens or maybe even a hundred times more. The bottom line is that our Solar System is a big place, and there millions upon millions of space rocks out there. There is not just a handful space pebbles out there. Instead, there is a vast array to choose from. Some of the options include space rocks that contain perhaps many billions of dollars' worth of platinum, other valuable metals and rare earth components.

On one hand, these space rocks could provide us key resources we are short on when we run low. On the other hand the bigger-sized space rocks are potentially severe hazards. If they were big enough and traveling fast enough, they could take out the entire human race, just like the 5–6-km wide asteroid did in destroying the dinosaurs about 66 million years ago. Bottom line, space mining could be good not only for replacing key resources but also could lead to technology that could help to protect us from being clobbered by a giant space rock in the future.

Technology Spinoffs

If the New Space economy is to provide a truly meaningful boost to the global economy and create an increase in new jobs, then one of the primary keys will be technological spinoffs. We know from the experience of the space programs in the United States, Russia, Europe, Japan, India and China plus other countries over the past 50 years that space-related research and development leads to the development of new and improved materials, new communications, robotics and computer capabilities, new products, new agricultural techniques, and new services. It is not possible to cite a major industry that has not benefited from space technology and systems spinoffs. Benefits have been derived from mining to construction, from transportation to banking, from health care to transportation, from education and training to energy companies. Batteries, laser systems, computers, air conditioning and heating systems, antibiotics, energy transmission and generation systems, building materials, trucks, automobiles, buses, trains, ships and aircraft have all been improved. These systems are better, faster, safer, and more durable due to space R&D programs. In the chapters that follow the new space industries will be described in terms of critical new technologies that will be needed to make new space systems viable. The following chart is indicative of the new technologies or systems capabilities that the new space industries will likely give rise to in the years to come. In many, many cases the new capability will be utilized here on Earth before it is actually used in space, or the so-called "Protozone" just below outer space (Table 2.1).

Table 2.1 Challenging new space enterprises and how they could impact the global economy

New Space Industry	Key New Technology, System or Application	Implications for the Global Economy and Future Employment
Space mining	• Improved remote robotics • Improved and cleaner mining & extraction techniques • Improved remote energy systems (solar cells-P/V cells, quantum dot energy systems, batteries, etc. • More efficient electric & ion thrusters & launch systems • More effective communications and IT control systems	Enhancements to the technical and operational capabilities of manufacturing & mining companies, energy companies, robotic manufacturing companies, and air transportation companies

(continued)

Table 2.1 (continued)

New Space Industry	Key New Technology, System or Application	Implications for the Global Economy and Future Employment
Solar power satellites	• Improved remote energy systems (solar cells-P/V cells, quantum dot energy systems, batteries, etc.) • Improved remote robotics • More efficient electric & ion thrusters & launch systems • More efficient long distance energy transmission systems	Innovations and efficiency gains for energy and power generation and transmission companies, improvements in the design & manufacture of solar power systems
Spaceplanes, Space Adventures	• Improvements in space range control plus tracking and guidance systems safety • Improvements in supersonic & hypersonic air transport propulsion and safety • Improvements in avionics and automated guidance systems	Improvements in the development of new supersonic and hypersonic aircraft, improved safety systems for all types of aircraft, improved air traffic management and control systems
Space habitats and space colonies	• Improvements in construction industries • New construction materials • Improved heating and cooling systems • Improvement in hydroponics and agricultural techniques • Improvements in remote health care and surgery	New materials for building construction, maintenance and refurbishment, improved and more energy efficient HVAC systems, enhanced district energy systems, improved farming techniques and genetic engineering of new agricultural products, range of improvements in health and medical care in remote areas
Space defense, planetary defense and traffic and control systems	• Improved radar and guidance systems • Improved broadband communications and IT networking systems • Directed energy systems • High-powered laser beams	Improved radar, remote guidance and tracking systems for all types of aircraft, improved broadband networking capabilities

The innovations discussed in the table above are, however, only the most obvious listings of the most predictable spinoffs that should occur as new technology, processes and techniques are developed to support the various New Space commercial activities that are now anticipated. Indeed the listings in this table are restricted to just the cutting-edge new space enterprises. Established space businesses such as satellite communications, remote sensing, space navigation, meteorological satellite services and on-orbit services will also likely have positive spinoffs of new technology and applications.

When the first transistor, solid-state computer, laser, artificial satellite, or synthetic material such as plastic were initially developed the actual practical uses of these new technologies or products were far from clear. In many instances some of the more important applications were totally unanticipated. The same is very likely to be true for the New Space industries that are burgeoning around the world and will come on line in the next 5, 10 or 20 years. This pattern of new and totally unique developments seems to especially percolate through such places as Silicon Valley. These innovation 'hot spots' seem to grow up in proximity to major research universities, governmental research laboratories and aerospace, computer and networking centers.

Where and How the New Gold Rush Will Begin

Certainly Silicon Valley seems to represent the almost ideal conjunction of interacting intellects that spawns innovation almost like spontaneous combustion. This rather uncanny place—as represented by Google, Facebook, Yahoo, Intel, a host of computer, communications, and genetic research companies, NASA Ames, various aerospace companies, Stanford University, and the Singularity University—seems almost unique, although there are a plethora of wannabes being cultivated around the world.

The students who convene at the highly selective Singularity University in Mountain View, California, on the campus of NASA Ames are perhaps indicative of those young aspirants that want to change the world. They are given the assignment of conceiving of a project that in a decade can have a positive effect on the lives of millions, if not a billion people. This never fails to get the creative juices flowing.

In such cauldrons of invention, such as Silicon Valley—and like-minded centers of learning and creativity—there is a strong likelihood that when one breakthrough innovation comes, it will trigger an avalanche of new findings and applications. Aerospace companies, telecommunications and networking labs, computer systems, artificial intelligence, and genetic engineering all churn around in proximity to one another, and new ideas flow. This "noosphere," a term coined by the philosopher Teilhard de Chardin in the 1900s, where there is a growth and sharing of information, ideas, and knowledge by creative minds on a global scale, brews a hardy stew of innovation that fuels invention and entrepreneurial startups. When ignited by the challenge of doing things in outer space, especially things that have never been done before, this powerful mix creates an intoxicating drive to change the world.

And there is enormous impatience to make it happen. New Space commerce has triggered new ideas, new companies, new technology, and an array of change with an urgency that is faster than ever before. The twenty first century is seeing the coming together of "future compression" time scales and a mushrooming of sometimes outlandish goals to go where no one has gone before.

It is these new "commercial space frontiers"—to accomplish the impossible—that start the dreamers dreaming. Space mining, unlimited clean power from the Sun, travel into the dark sky of space, self-sustaining colonies within the reach of Earth's gravity—these are the ideas that motivate remarkable people such as Elon Musk, Peter Diamandis, Bob Richards, Jim Keravala, Rich Tumlinson, Paul Allen, Robert Bigelow, Jeff Bezos and Sir Richard Branson. About half of these powerhouses of innovation the author has gotten to know through the International Space University, the Singularity University or at space conferences. When one sees these mere mortals up close you sometimes overlook the magnitude of their daring dreams and their commitment to change the world. But change the world they aspire to do.

The emotions they engender are two-fold. Awe and inspiration is the first reaction as to the magnitude of their goals. This is followed by concern and caution as to whether the world is ready to change so rapidly. With a world filled with tension and jihadist extremism and political leadership that sometimes has difficulty recognizing global challenges such as climate change, over generation of greenhouse gases, and the threat of global pandemics, it does at times seem doubtful that human are fit to colonize the world—let alone the universe. This is why we must look to innovation that produces more than neat new products. Rather, we need inventions that can usher in a sustainable world that can: (1) survive over the long term; (2) curb population growth; (3) figure out the twenty first century human employment conundrum; and (4) create new economic systems that usher in a better and more productive future.

In the latter chapters of this book we indeed look back to these very real concerns. In doing so we look to what types of rules and regulations might deliver us into a better world and new type of space economy. In a McGill University landmark study of the "Global Governance of Outer Space" efforts have been made to see where new standards, laws or policies could help. This 2-year effort has drawn on the expertise of over one hundred space lawyers, regulators, space scientists and engineers [4]. The purpose of this global and comprehensive study of the changing world of space was to explore what changes we need to make in terms of national legislation, codes of conduct, and international regulations to get ready for the new gold rush. This effort is a prelude to the discussions that will take place at the Unispace + 50 Conference to be held in Vienna, Austria, under the auspices of the UN Committee on the Peaceful Uses of Outer Space in 2018.

Conclusions

Some may doubt that New Space commerce is the vital threshold to the future. These skeptics see important research and innovation occurring in biological, chemical, computer networking, artificial intelligence and energy research, among other scientific fields of inquiry and dismiss the need for off-world enterprise as a 'secret sauce' ingredient that serves as the top fuel to major new invention. Certainly we have other challenges to motivate and vex us. These include such challenges as climate change, addressing oceanic pollution, changes to the icecap albedo, overpopulation and underemployment, new epidemics, plus global hunger, health care and educational needs. Certainly these challenges could and should also spur us to innovate. But it turns out that space technologies are often the vital means that allows us to cope with most of these challenges from a fresh new perspective.

Remote sensing and meteorological satellites that monitor the entire planet synoptically are vital to measuring climate change and determining pollution levels on the land, the seas, and arctic regions. These eyes in the sky are key to monitoring crops, disease in trees and vegetation and reporting on patterns of human settlement. Communications satellites are key to tele-education and tele-health, and broadband networking around the world. These links in the skies also supports rescue operations, banking, and all sorts of transportation systems. Space navigation satellites are vital to transportation safety, food and drug shipments and distribution, and even the synchronization of the Internet. The original meaning for what we now identify as satellites was provided by Galileo. That word was actually the Latin word *satelles,* and it meant helper or servant. The truth of the matter is that we are today much more dependent on our space servants that most of us ever suspect. If you are just a bit curious about the extent of this space-based dependency go to "You Tube" and find the short video entitled "If there were a day without satellites."

The truth of the matter is that satellite-based communications, networking, monitoring, transportation routing, positioning, navigation, weather forecasting, rescue, and safety systems are now vital to our everyday existence. One of the great lures of outer space is to explore the unknown riches of the cosmos. In a wide variety of ways space applications, space science, space transportation and space exploration will either guide our future or provide the technology to allow humans to have a future. If our technologies fail us, human civilization as we know it may also fail.

There are those that say "Why waste money on outer space?" To those critics we point out the following: The money that is spent on space is a part of the global economy. The equipment that is developed and the salaries paid in truth benefit real human beings. Satellites provide news and entertainment to billions of people. The Internet, aircraft takeoffs and landings, automobile navigation systems, global shipments and national defense systems—all this and more depend on space navigation systems. The warnings against violent storms and hurricanes, the knowledge about how to combat climate change, and rescue systems for stranded pilots, seaman and passengers—these, too, all depend on our servants in the skies. Much of our knowledge of the universe, and clear warnings of cosmic dangers from the Sun and asteroids, depend on satellite servants in the sky.

Had the dinosaurs had a space program their kind might have survived the asteroid hit that exterminated them. Let's hope that we humans are smart enough to invest in space to not only colonize the cosmos and learn from whence we came but also to survive as a species. As far as we know not one red cent spent on space has ever gone to pay an alien. Space is actually a very human enterprise. And indeed it is the most likely enterprise to open a totally new window on the future.

References

1. Joseph N. Pelton and Peter Marshall, MegaCruch: Ten Survival Strategies for the 21st Century (2013) The Emerald Planet, Washington, D. C.
2. Ray Kurzweil, *How to Create a Mind* (2012), Viking Press, New York.
3. IBM Watson Health: Welcome to the New Era of Cognitive Healthcare (2016). www.ibm.com/smarterplanet/us/en/ibmwatson/health/. Last accessed May 21, 2016.
4. Ram Jakhu and Joseph Pelton, *Global Governance of Outer Space* (2017), Springer Press, New York.

3

The Expanding Use of Space in Communications, Navigation, Remote Sensing and Weather Satellites

The idea of a new gold rush may seem farfetched to some, but there is hard data to support a growing—even exploding—space economy that is happening today.

There is a wealth of "electronic space riches" already flowing from space back to Earth today. In case you have not been watching lately space commerce is a part of today's global economy and touches every aspect of our daily lives. Operational satellite networks are a part of our vital global infrastructure, and if they were suddenly to be wiped out it would literally cripple the world as we know it. As many as 500 million new people will log on to the Internet for the first time in the coming year, and many of these new Internet users will directly or indirectly log on using a satellite connection.

There are today over 1250 operational satellites, with a net value that exceeds $2 trillion (U. S. dollars). These satellites provide vital services, with only about 5 % of those engaged in purely scientific missions. Fig. 3.1 below shows that communications, networking, broadcasting, navigation, remote sensing and surveillance are the main applications for these servants in the sky. Direct revenues from commercial satellite applications, launch services and manufacturing, according to the Tauri Group Report for the Satellite Industry Association (SIA) totaled over $200 billion (U. S. dollars). What is particularly notable about this number is that total revenues—using the same SIA reporting system—were only about $90 billion in 2005. In fact, there has been steady and ever-increasing growth in commercial satellite operations for every year over the past decade—even in years of global recession. Few other multi-billion dollar industries can make a similar claim. If the same trend continues through 2025, the total revenues could reach a half trillion dollars a year by then.

© Springer International Publishing Switzerland 2017
J.N. Pelton, *The New Gold Rush*, DOI 10.1007/978-3-319-39273-8_3

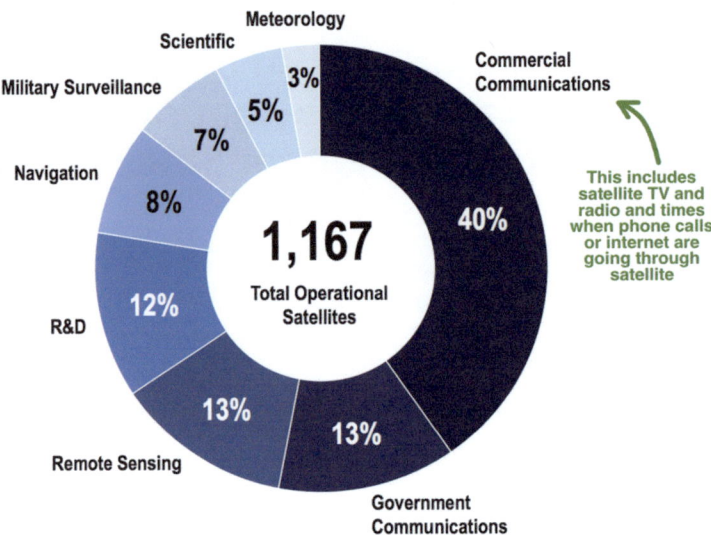

Fig. 3.1 A breakdown of operational satellites by function (Chart from the Satellite Industry Association Report for 2014.)

One might say well over $200 billion—or even $500 billion—is not much in terms of a global economy that is reckoned in many trillions of dollars. But this is only the beginning of the story.

First of all, there is no way to place a commercial value on military satellite functions or meteorological satellite functions. In terms of public safety and national defense the value of these various space assets are presumably very high. If one were to try to estimate this value in some sort of a strategic and social benefit valuation, the $200 billion number attributed to commercial satellite applications might well double.

Further there is no lessening of growth in sight. A new wave of expansion is now underway. First of all there are now plans afoot to deploy large-scale low Earth orbit satellites optimized for Internet traffic. These new systems, with names such as One Web, LeoSat, and others just being conceived could lead to a new wave of satellite growth. In addition there are also major innovations in what might be called "conventional" communications satellite systems deployed in geosynchronous orbit. These are called "high throughput satellites," and they are also serving to expand space services and revenues.

The story does not end here. The derived "value added" economic results from these various satellite applications are truly enormous. The next generation ("next gen") air traffic management and control system is satellite-based, and many aircraft take-off and landing systems today are already dependent on the GPS or other satellite navigational services. The same is true for global shipping operations, surveying operations, and hundreds of other commercial applications. The ships at sea are, of course, not taking off, but they are being routed more efficiently. Thus large super tankers and freight-carrying ships get to port more quickly. The resulting savings total in the millions, if not billions, of dollars by cutting days off of cross ocean trips and by avoiding dangerous storms.

The derivative value of space-based precise timing and navigation applications for air, sea, and ground transportation, for security systems, and for scores of other commercial applications today has been placed at numbers as high as $150 to $200 billion. Or if one were to look to communications and broadcasting satellites and the derived value in terms of improved efficiencies in airline bookings, routing, and safety, banking operations, stock trading, credit card validations, insurance company operations, news and entertainment broadcasting radio and television, and on and on. The economic impact numbers reaches into the many trillions of dollars.

Or, one could look at the equation upside down. If the GPS and other satellite navigation satellite networks were to suddenly not be available we would lose the synchronization of the Internet in many countries. Airline operations would be more dangerous. Satellite navigation will be key to supporting the operation of new driverless cars. The truth is that if we were to lose our navigation and timing satellites this would in a short order of time literally cripple the global economy. The impact would be felt everywhere. The losses would be felt in modern military systems, airline networks, information technology networks, banking and commercial retail authentication systems, and other vital infrastructure would be stripped of its core functionality.

Today's modern society satellite networks have become like electric motors and solid state digital processors. They are very prevalent but hidden from view in the skies, so that people do not realize they are there. There is the probably apocryphal story about the U. S. Congressman who, at a hearing on appropriations for a new meteorological satellite, was supposed to have said: "Why do we have to spend all these millions on weather satellites when I can just turn on the television news each morning and see what the weather is going to be?" When communications satellites first came on the scene it was a big deal, and the Olympics in 1968 from Mexico City and 1972 from Montreal consistently displayed a message at the bottom saying "Live via Satellite." Today people watch some 20,000 satellite video channels around the world without knowing or caring how they get their global news, sports, and entertainment. They simply rely on these technological marvels in the skies and demand ever more space-based communications connectivity without realizing the amazing innovation that underlies these machines in the skies.

The New Guys on the Block

And why is it a very good bet that commercial satellite industry revenues will reach the half trillion dollar level in a decade? One very good reason is that the current market is churning faster than ever before, with change and innovation as noted above.

Today the satellite communications market is reeling from a salvo of new technologies and bold new entrepreneurs and engineers that are re-imagining what was until recently seen as an established and mature market. Mark Dankberg, with his Via Sat spacecraft, and Pradman Kaul of Hughes Network Systems, with his Jupiter Technology spacecraft, among others in the communications satellite game, are now deploying new high throughput satellites that have the ability to handle upwards of 100 Gb/s. ViaSat 1 and 2, Echostar XVII with Jupiter Technology on board as well as Intelsat's Epic Satellites are almost like multi-destination fiber optic cable systems in the sky. These hugely capable satellites have tens of thousands of times the throughput capacity and the cost efficiency of the Early Bird satellite that started commercial satellite services a half century ago, in 1965 (see Figs. 3.2 and 3.3).

And progress in satellite technology does not stop there. Three Silicon Valley wizards, named Greg Wyler, Brian Holz and David Bettinger, left Google in 2013 to create a company called WorldVu. This company, now known as OneWeb Ltd., has managed to receive backing from the Virgin Group, Intelsat, Qualcomm, Air Bus, HNS and SpaceX to create a network of some 700 small satellites—including spares—to be deployed in low Earth orbit. The idea is to create a huge satellite constellation that is optimized for broadband Internet networking and particularly geared to serve developing countries.

Greg Wyler, who waxes eloquent about the possibilities, told of his ambitious plans during a dinner in his honor when he received the Arthur C. Clarke innovation award. He sees a whole new ballgame for global satellite communications. His plan is to deploy in low Earth orbit about 500 miles (800 km) in altitude some 700 quite small satellites, including spares. These small satellites will weigh only 125 kg (275 lbs.) (see Fig. 3.4).

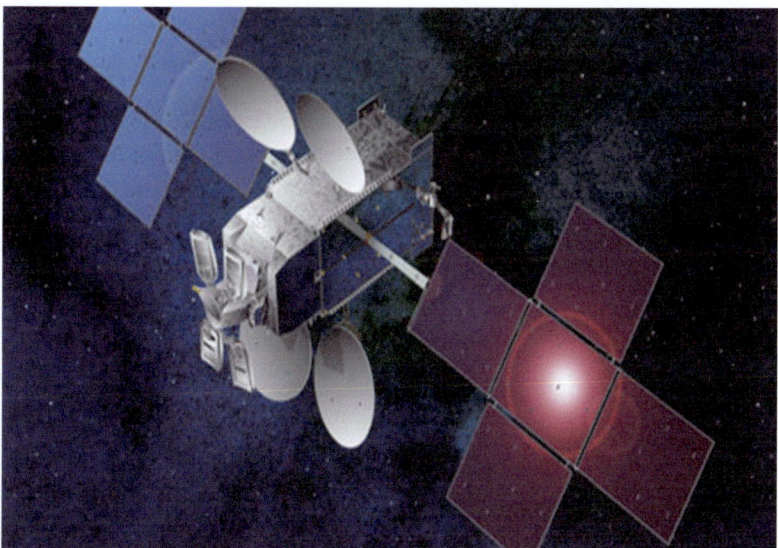

Fig. 3.2 The Echostar XVII represents one of today's high throughput satellites (Image courtesy of Hughes Network Systems.)

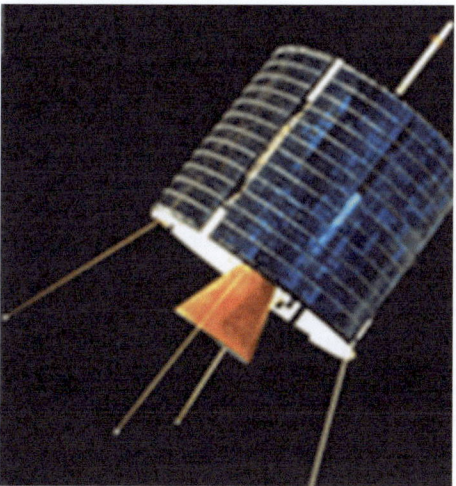

Fig. 3.3 The tiny 40-kg Early Bird satellite launched in 1965 started it all (Image courtesy of Comsat Legacy Project.)

The objective is to churn out these small satellites like video cassette recorders or television sets and then quickly launch them at low cost via Virgin Galactic (Launcher One) and SpaceX (Falcon 9) launchers. These are the new commercial launcher systems backed by space billionaires Sir Richard

Fig. 3.4 The OneWeb 700 satellite low Earth orbit constellation planned for 2017 and 2018 (Image courtesy of OneWeb.)

Branson and Elon Musk. Who knows where this all stops and whether the feisty "space labs" with their new small and even lower costs rockets will shake up the satellite launching business even further.

Greg Wyler's first ambition was to bring low cost broadband communications to the country of Rwanda using terrestrial technology. He launched a company called Terracom to lay fiber optic cable in Rwanda. But he soon found he could not make international networking for Africa viably possible using telecom systems on the ground. The only systems that seemed to ultimately make sense to Wyler after study were satellite networks. He quickly twigged to the fact that a satellite constellation could cover the entire developing world where over three billion people live. Instead of building an Internet system for a few million Rwandans, he leaped to the idea of service for what he affectionately called "The Other Three Billion." He started out with a medium Earth orbit constellation with only 12 satellites covering the equatorial region and called it O3b (for the Other Three Billion), and then moved on to launch planning for the Mega LEO OneWeb system [1].

This revolution in truly low cost satellites deployed as large-scale constellations in low Earth orbit, however, did not start with OneWeb. No, it was truly started in concept by a firm originally called Skybox. This project was started by four graduates of Stanford University in Silicon Valley in 2009. They had the idea of using commercial off-the-self (COTS) components to create a remote-sensing satellite network on the cheap but with the ability to get rapidly updated information. They built the satellites with a team of young graduates that looked like teenagers and lived in what looked and felt like a frat house. This project helped start the New Space revolution in how people thought about space.

Skybox was such a success that it was gobbled up by Google for some $500 million in a deal that was put together only 5 years later in 2014. Today's constellation provides rapidly updated remote sensing and now supports Google Mapping. As of March 2016 Google announced that Skybox's new name would become "Terra Bella," although many feel the original name captured the spirit of this brash start up much better than the new name. The four originators who garnered more than $100 million each are not quibbling about the name change as they take their amazing financial coup to the bank [2].

This brash new satellite system now has many imitators. Google is squarely in the "disruptive technology" business, and it sees "New Space" systems as one of the new ways to change the world. Disruptive technologies are ones that disrupt and replace established technologies.

Currently Google, Facebook, SpaceX and Amazon.com have all indicated various plans or concepts for new commercial space businesses focused on creating an Internet in the skies. These initiatives include new cost efficient launch systems, 3D manufacture of new types of satellite networking constellations and even high altitude balloons to support Internet connectivity. Silicon Valley has apparently discovered the "New Space religion" and how it can change the world. Space commerce now represents a part of their growth strategy. With very moxie corporate titans who embrace disruptive technologies—such as Larry Page, Mark Zuckerberg, Elon Musk, and Jeff Bezos—the world of space will apparently change at an even faster pace. Even Paul Allen of Microsoft fame is in on the act via his Vulcan Technologies that is now leading the development of the huge Stratolaunch vehicle that will serve as a high altitude carrier plane for new types of commercial launch vehicles. These space billionaires, one and all, are jumping into the space systems manufacturing and commercial launch business. The winds of change have suddenly reached new altitudes.

Sorting Out Today's Commercial Space Enterprises

There was a time when one could refer to the satellite industry and mean the communications satellite industry. But today there are a myriad of new and prospering commercial and satellite-related industries that are sufficiently well established and unique that it makes sense to explore these one at a time in terms of what businesses they do and their growth potential.

The Communications Satellite Industry

Today over 53% of operational satellites are for communications, networking or broadcasting services, if we take into account all of the commercial communications satellites (40% of the total) and the military and governmental communications satellites (13% of the total). And there appears to be substantial growth ahead. With the current plans of new business entities such as OneWeb, which intends to deploy the so-called mega-LEO (or large-scale low Earth orbit satellites), the total number and the percentages related to communications satellites could increase substantially. The communications satellite industry, however, is today divided into a number of key parts.

The biggest sector in terms of revenues is the satellite broadcasting industry. Companies in this field broadcast television directly to homes and businesses and in the process sell video entertainment directly to consumers on a retail basis. Because of their vertical integration, which includes satellite deployment and operation, entertainment sales to the home or business client, etc., they have by far the largest revenue stream and greatest profitability. Most of the other satellite service providers are one or two steps removed from the consumer and thus have lesser revenues and lower profit margins. The categories of communications satellite services thus include: (1) broadcasting and direct to the home television and radio services; (2) so-called fixed satellite services; (3) mobile satellite services; and (4) business-to-business (B2B) data relay.

TELEVISION BROADCASTING AND DIRECT-TO-THE-HOME SATELLITE SERVICES. The biggest players in this market are the two U. S. television satellite service providers. These are DirecTV (now a part of AT&T) and Echostar/Dish, both operated from the Denver, Colorado, area. There are many ways to make a profit from these satellite-based video services that includes a wide range of paid subscription services for movies (Starz, HBO, Cinemax, etc.), pornography (i.e., *Playboy*), sports, gambling, and on-line shopping. In addition these satellite networks can offer software download services as well as music. The European market is split more widely with Eutelsat, BSkyB, and SES vying for market share. Mexico, Japan, Korea, Canada, Australia, China, Brazil, India, the Middle East, South America, Africa and the rest of Asia all have satellite broadcasting and distributed television services. Altogether this is big business, with new types of business such as the downloading of video games and providing broadband digital services to businesses on video transponders, all helping to create new growth opportunities. The satellites can also be used to provide broadband Internet services as a downstream system that can be linked to terrestrial telephone lines for the upstream connection. In rural and remote areas this is sometimes the only available option.

A big change occurred when digital video compression technology suddenly allowed up to 18 video channels to flow through a single satellite transponder, where in the past only one or two video channels could be sent downstream. In this new digital satellite environment the number of video channels provided worldwide has expanded like gangbusters. It is estimated today that there are 20,000–30,000 video channels operating via satellite around the world. Some of these video channels generate millions of dollars of revenues (particularly the most popular movie channels, plus porn, sports and gambling channels).

Some of the newer sports and gambling channels are very, very profitable. Most of the broadcasting satellites operate in a band designed for direct connection to the home or business, but some satellite networks that operate in the lower band for so-called fixed satellite services provide what is called a direct-to-the-home television service using so-called backyard dishes. The latest version of these "conventional" satellites are powerful enough that they can provide service to antennas that are only a meter or so in diameter, and many of them offer services to apartment complexes. The European-based SES began by offering their Astra service subscription television services and was the first entity to make this a major enterprise.

In the United States the "backyard dish" market began with rural and remote home owners getting "descramblers" to obtain video services like HBO and Cinemax for "free." Today this U. S. market has now firmly entered the age of digital encryption. Thus this service has virtually all moved back to the direct broadcast satellite service providers, namely DirecTV and Dish/Echostar. In the age of cable television and Internet-based video services, however, competition is now everywhere.

DIRECT AUDIO BROADCASTING SERVICE (DABS). One of the newer broadcast satellite services is audio broadcasting. XM Radio and Sirius Radio (that have now merged into a single company) were the first of these services, and they targeted the U. S. market. The business plan for XM/Sirius is dominated by the concept of providing high quality music and audio entertainment to automobiles in the U. S. market. The initial rollout of these satellite services (that operate on a lower frequency than the television broadcasting satellites) were tied to the various U. S. car manufacturers, i.e., General Motors, Ford and Chrysler.

DABS-based offerings include not only "coast to coast" high quality music but comedy, variety, news, and—under an incredibly high-priced contract—the radio host Howard Stern's rather smutty adult entertainment show. Howard Stern today boasts an annual salary of $95 million and a net worth of some $600 million. These radio satellite offerings become a paid monthly service after the free service runs out under a new car purchase package.

Perhaps the most well-known and widely advertised DABS is GM's "OnStar" service, which includes not only news and entertainment but other important offerings, such as an anti-theft tracking system, an automated unlocking service if your car is locked with your key inside, as well as roadside assistance and communications if someone is involved in an accident. The OnStar satellite service can now be upgraded to a 4G LTE cellular service as well. Ford and Chrysler offer similar satellite services with parallel features now that XM and Sirius have become a single entity. All of the automotive companies are adding communications and IT service capabilities to their products in order to create ongoing revenue streams to supplement the revenues derived from car and truck sales.

There were plans to provide a parallel service in Europe, but it never got off the ground. Another very ambitious project to provide a DABS service to the world started with an offering called "Africastar." This was not only to provide channels for commercial radio stations that would broadcast to all of Africa and parts of Europe and the Middle East. It was also to provide educational broadcast services and other features. It turns out that the cost of the ground radio receivers were too high—especially when African nations put very high tariffs on these satellite radio receivers. The builder assumed ownership of the "Africastar" satellite when it ran into financial difficulties, and the other satellites for the rest of the world were not deployed.

Fixed Satellite Services (FSS)

The first commercial satellite system was envisioned as a single integrated global system that would provide new international links across the oceans. As of the 1950s and 1960s there were coaxial submarine telephone cables that were able to provide international telephone connections. But the largest submarine cable had a capacity of 36 voice circuits. The technology then known as Time Assignment Speech Interpolation (TASI) was able to double this capacity by assigning channels only when people were talking. This created a quite annoying clipping effect.

When the new global Intelsat Consortium launched the quasi-experimental Intelsat I, known affectionately as "Early Bird" in April 1965, international telephone connection capacity suddenly jumped to 240 voice channels or one low-quality black and white television channel. This newly operational geosynchronous satellite—a minnow in comparison to today's Comsat whales—was hailed as a major technological achievement. Although today's high throughput satellites last ten times longer and have 10,000 times more capacity, Early Bird represented a true breakthrough in global international communications.

By 1971, just 6 years later, Intelsat IV was launched with 4000 telephone circuits plus 2-color television channels, and the satellites have just kept getting larger and larger, and more and more powerful. This increase in performance exactly traced the famous "Moore's Law" for computer progress. Moore's original projection stated that computer chip performance would double every 18 months. In the case of satellites it was a 16-fold increase in capacity in 6 years.

And this progress has continued. The satellite world has kept moving along in its exponential race forward into the future. This is really not too surprising in that satellites today are simply sophisticated digital processors in the skies with specialized software that defines their function and continues to be improved.

The continued growth in performance, power and capacity of the satellites in the skies allowed the ground receivers to shrink ever smaller in size and lower in cost. Earth stations that once were virtually ultra sensitive radio telescopes that cost many millions of dollars and were 10–20-ton monsters had to be staffed 24 h a day by 50–60 people. Today satellite receivers are called very small aperture terminals (VSATs) or even ultra-small aperture terminals (USATs). These low cost satellite receivers are now a meter or so in diameter. Indeed for mobile satellite applications the user transceivers can even be handheld gadgets only slightly bigger than cell phones. These amazing satellite telephones are shown in Fig. 3.7 later in this chapter.

The highest capacity satellites today could transmit the equivalent of 25,000 television channels, or 25 million telephone channels. Only computers and telecommunications devices have seen this type of exponential growth for decade after decade.

The initial satellite communications service that the Early Bird satellite represented was designed to connect fixed Earth stations around the world that had a clear path connection to the satellite. But the only constant in the satellite industry has been frequent change—change of satellite and ground system technology and change of service and applications.

Quickly FSS satellites changed to include all types of services such as national long distance telecommunications, forms of television and radio distribution services, data networking, credit authentication, and provision of enterprise networks to large corporations. It also provided connectivity for people in rural and remote areas. This rapid expansion of services gobbled up the radio frequency spectrum allocated to this service. First the C-band frequencies were saturated, next came the higher Ku-band-frequencies and now the even higher Ka-band frequencies are being consumed. Thus, even higher frequencies in the so-called millimeter wave bands (i.e., the Q, V and W-bands) are likely to be needed to meet future growth. These expanded uses of frequency bands plus the even more important digital compression techniques and the ability to launch more and bigger FSS satellites has allowed this enormous growth.

Satellite networks have expanded to serve more and more of the world, then expanded again and again. Wireless satellite services and broadband wireless cellular services on the ground have an insatiable appetite for frequency bands, and today unfortunately they are in competition with each other. People want satellite service and they want broadband cellular and this becomes a headache for national and international regulatory authorities. Sometimes a difficult decision has to be made. Do we allocate frequencies for safe takeoff and landing of aircraft or international banking, or rescue of pilots and ships, versus providing more broadband cellular for people to watch football games while sitting in the stands? The engineers are tasked with finding more efficient and digitally compressed use of frequency spectrum so that all the competing needs for radio frequencies can be met. Through the use of multiple cellular beams both satellites and terrestrial cellular systems and digital multiplexing the radio frequencies are a 1000 times more efficient than they were 30 years ago.

When FSS satellites first began an equivalent telephone circuit for 1 year was quite expensive and cost on the order of $64,000. Today the wholesale cost of service on so-called high throughput satellites is on the order of only $5 to $10 per annum. There are those that suggest that fiber optic networks can provide service at an even lower cost, on the order of just $1 a year. But the costs are now so low and satellites and fiber are so cost efficient that one could say their costs have become almost irrelevant. In short the cost of sales, advertising, billing, operations, and management in telecommunications and networking outweigh the cost of either a satellite or fiber optic telephone or data link in terms of the capital investment. The issue thus becomes one of service availability and flexibility to meet customer needs.

This means that fiber is more efficient for connecting large pathways between cities or broadband links to homes and businesses in urban areas that are densely settled. Satellite networks on the other hand are best for broadcasting, multi-casting and mobile application (Table 3.1).

Table 3.1 Satellites versus fiber optic networks

Areas Where Satellites Can Have an Advantage	Areas Where Fiber Optics Can Have an Advantage
(i) **Broadcasting operations:** Television or radio operations that are providing service to many customers. The truth is that cable television stations get their programming at their head ends by satellite distribution. HBO, Cinemax, Starrz, CNN, etc., all start out as satellite television distribution. Over a billion people get their programming via satellites.	(i) *High-density "trunking routes" between cities and countries:* Fiber optic cables are free from atmospheric interference and are very cost effective for high volumes of telecommunications and IT traffic.
(ii) *Large-scale networks that are constantly changing.* This includes retail outlets with thousands of nodes that are constantly changing such as service stations, car dealerships, ATM machines and retail stores that need credit card authentication.	(ii) *Cable television networks:* As cable television systems have added more and more channels and support more high data rate IT services, they have increasingly converted to fiber optic networks, although some are still using coaxial cable.
(iii) *Rural and remote locations:* Services such as to islands, farms, mining and oil rig operations and scientific research centers in remote locations.	(iii) *In-building networks for television and high data rate services:* The newer office buildings and apartment complexes are installing fiber for conduit risers. They may use Wi-Fi networks to operate off these fiber cable installations.
(iv) *Enterprise and multicasting applications:* This is for downloads of new computer software or movies or business video distribution to branch offices with narrow band requests for information, updates with regard to inventory, billing, etc. This type of asynchronous service works well for enterprise networks for Ford, 7–11, etc.	(iv) *Any application that requires high throughput and is in close proximity:* This would include local area networks (LANs), wide area networks (WANs), and metropolitan-wide area networks (MANs).
(v) **Users requiring mobile access:** The big advantage of satellites over fiber comes with broadcasting and when mobile access is required.	

Mobile Satellite Services

The third key satellite service that is now a fast-growing space application is the so-called mobile satellite service (MSS). The satellites of the 1960s were small, had very little power and could only connect radio signals between giant Earth stations that were located in very remote locations that were isolated from other types of radio wave interference. The signals were so faint they were at a power level equivalent to a snowflake falling to Earth. Clearly in this technical environment, mobile satellite communications were simply not feasible.

The communications satellites of the 1970s and 1980s became more and more powerful, and 3-axis body stabilized platforms allowed large antennas to concentrate and transmit power to specific locations. The technical result was the ability to concentrate radio concentrations in powerful ways so that the ground antennas could shrink in size and complexity. The practical and economic consequence is that mobile satellite service—at least at the level of ship-mounted antennas—became feasible. Actually this was first accomplished to accommodate NASA communications in support of the Project Mercury and Gemini flights, where a very large antenna was put on a ship to communicate with an Intelsat II spacecraft, but this was far from a cost-effective commercial operation.

In the years that followed Comsat, a global telecommunications company based in the United States, deployed the Marisat satellite that proved this type of regular communications links to ship-mounted dishes equipped to track a satellite even at sea were possible on a viable commercial basis. This satellite was utilized to support both the U. S. Navy and commercial operations. This groundbreaking, or perhaps one should say "space-breaking," satellite lasted over a decade in service and set a new course that resulted in more and more sophisticated satellites that can provide not only maritime mobile satellite services but all sorts of services that include maritime, aeronautical and even land mobile services to ground units that are the size of a laptop computer or even a handheld unit (see Fig. 3.5).

Eventually the Marisat satellite, plus the Marecs designed by the European Space Agency and three specially equipped Intelsat V satellites with maritime satellite packages on board, were made available to cobble together a global satellite network for the new INMARSAT organization. Although three different satellites were utilized INMARSAT was able to provide routine communications to ships at sea. As the satellites became more and more capable and larger and larger, antennas were designed to unfurl in space and provide mobile satellite services that were more and more amazing. Today mesh antennas some 20 m (66 ft) to 26 m (86 ft) in diameter are being manufactured by the Harris Corporation to support commercial and defense communications satellite systems that are being deployed in geosynchronous orbit. One of these giant satellites in geo orbit with a giant antenna can provide service to almost one-third of the world's surface and support links to even handheld units (see Fig. 3.6).

3 The Expanding Use of Space in Communications... 53

Fig. 3.5 The world's first dedicated mobile communications satellite (Image courtesy of Comsat Legacy Project.)

However, there are other technical solutions to providing mobile satellite links. The alternative is to deploy a constellation of satellites much closer to Earth's surface. This is the approach taken by systems known as Iridium (66 satellites plus spares) in low Earth orbit or Globalstar (48 satellites plus spares) in a different configuration. These satellites that are 30–40 times closer to Earth do not have to be as large or have giant antennas to complete a link to users on the ground.

In either case the approach taken for all these mobile satellite systems is what might be called technology inversion. The satellites or the constellation is powerful, complex and support highly focused beams so that spacecraft can "talk" to a very small handheld device. The prediction made by Arthur C. Clarke in the 1950s that 1 day people could talk to anyone in the world via a small handheld device has become reality a half century later (see Fig. 3.7).

Today there are a number of mobile satellite networks that can allow communications to handheld devices that cost on the order of $1000 a pop. These systems include Inmarsat, Thuraya, Iridium, Globalstar, Skyterra, and

Fig. 3.6 Harris Corporation testing a new large aperture antenna design for a mobile satellite (Image courtesy of the Harris Corporation.)

Terrestar. Iridium and Globalstar only provide compressed telephone and limited data at 2.4–4.8 kb/s. The others provide video and broader band services in the 400 kb/s range.

To state that the current offerings are fluid and difficult to define with precision is actually a very excellent summary. Iridium is deploying a new and more capable Generation NEXT in low Earth orbit. Globalstar is going from a low Earth orbit to a geo orbit system. Skyterra that was deployed by a company named LightSquared is going through bankruptcy proceedings and selling their satellite. Inmarsat is deploying a new Express system that is offering the ability to provide mobile- and land-based fixed satellite services. Satellite companies such as Intelsat that have traditionally offered so-called "fixed satellite services" are offering broadband video and telecommunica-

Fig. 3.7 Mobile satellite handsets available from Globalstar, Inmarsat and Iridium, respectively (Images courtesy of Globalstar, Imarsat and Iridium.)

tions services to ships at seas while Inmarsat is offering not only land-based mobile services but also video links to broadcasters and military units in stable locations. Everybody is competing with everybody, and the huge new capacities that come with the new high throughput satellites is serving to heat up the competition with lower prices favoring buyers in the global market. The new entrants from the disruptive technology-oriented players such Google, Facebook and Amazon.com are just making the competitive juices flow even faster.

The bottom line is that everything is in flux. The old divisions between broadcasting satellite services (BSS), fixed satellite services (FSS) and mobile satellite services (FSS) under which the International Telecommunication Union (ITU) allocates frequencies are not being strictly enforced nor observed by anyone anymore. The main concern at the ITU and national frequency regulatory agencies is to avoid direct interference between and among these various systems and to try to at least keep commercial communications satellite services and military communications satellite services sorted out. This, too, is very difficult because military and defense agencies are buying many of their services from the commercial world due to the fact that their costs and services are much less than the governmental or military communications satellite systems.

The ITU has no "frequency police," and it is simply seeking to avoid major interference or jamming of the various satellites that would prevent services from going through. Today the largest instances of jamming and interference are coming not from overlapping commercial systems but rather those originating from Iran and Syria. This has proven to be a very large challenge for ITU officials in Geneva, who have gone to the governments of Iran and Syria, saying that they are finding evidence of jamming of satellite services coming from within that country. Would you please investigate and find the jammers and get them to stop? The authorities in these countries say, "We will investigate this problem and get back to you." They then report back. "Sorry we have not been able to find those that are carrying out this interference with the satellite service." The ITU is not able to go into these countries and find the sources that are causing the interference.

Data Relay and B2B Satellite Services

There is one other type of commercial satellite service provider, those that supply data relay or business to business (B2B) satellite links. These are systems such as Orbcom. These systems provide simple store and forward data messaging and often are also designed to provide GPS position location as an integrated service. These LEO constellations can be used to support shipping, car rental service auto tracking, and oil and gas pipeline monitoring. This low cost satellite service has provided an option for fleet operators that do not require broadband communication satellite services.

This type of satellite service that operates in lower VHF and UHF frequency bands is also employed on a non-commercial basis to relay data to remotely located doctors (i.e., Lifesat) and others that need global connectivity to rural and remote areas. The first to use small satellites for such data relay were the amateur radio ham radio operators with the Amsat Oscar satellites. The Surrey Space Centre in England has designed and helped to launch a number of small satellites for data relay.

Finally the space agencies around the world such as NASA, JAXA and ESA have a number of low Earth orbit (LEO), medium Earth orbit (MEO) and geosynchronous Earth orbit (GEO) satellites in operation that relay data from their lower Earth orbit constellations to processing centers in real time. These data relay satellites that connect from LEO satellite to GEO satellites and then back to data processing centers operate in scientific bands and are very expensive multi-billion-dollar satellite facilities. The NASA Tracking and Data Relay Satellite (TDRS) system was the first of these type of satellites.

Military and Governmental Communications Satellites

Essentially 13% of the 1200 operational satellites deployed in Earth orbit, or about 150 satellites, are either military communications satellites or governmental satellites such as the data relay satellites discussed above. This is a significant number of the satellites that are providing vital services around the world and include satellites for mobile communications services, UHF communications to ships, high data rate services, etc. In addition to dedicated military communications satellites that come equipped with shielding, radiation hardening, and special encryption there is significant use of commercial communications for so-called dual-use purposes. This means using conventional commercial satellites to support various types of military operations and can be as straightforward as sending television and radio entertainment services overseas for troops stationed abroad to supporting e-mail videoconference communications between soldiers and their families at home. In some instance they may be used for relaying UAV imaging back to a command center.

Recently there have been innovations that have created what might be called hybrid systems. The so-called XTAR system that was a joint venture between the aerospace company SSL and Spanish business and banking interests deployed a communications satellite that operates in the military X-band but is offered as a commercial service. In short, this is a commercial communications satellite that operates only in the military X-band and leases its transponders to the U. S. military or Spanish or South American military forces (Fig. 3.8).

Military forces in the U.K., France and Italy have also made the transition to having aerospace contractors design and build military communications systems that they lease on a long-term basis and then allow the contractor to sell or lease additional capacity on a commercial basis.

In summary the military communications satellite world is now quite complex. There are more and more options available, which include dedicated military satellites, some of which are dedicated to national military operations and some of which are shared among strategic national allies around the world. Next there are build-to-order military satellites systems leased from private contractors, commercial satellites engineered to operate only in military bands that then lease capacity to military users, and commercial satellite systems operating in commercial frequencies that lease capacity to military users on a "dual-use" basis.

This means that in addition to the 150 satellites deployed to support governmental research, operations (i.e., satellite links to embassies) and dedicated military communications there are a large number of commercial satellites

Fig. 3.8 The innovative XTAR satellite (X-band service only) that exclusively serves military networking requirements (Image courtesy of Loral Space and Communications Ltd.)

supporting military communications needs in one or more of the "dual-use" modes of operation outlined above. The U. S. Department of Defense has spent billions upon billions of dollars to create what is called the Global Information Grid (GIG) to create an intelligent telecommunications and command capability that encompasses the entire world, and dedicated and dual-use satellites are a critical component of this vast network of fiber optic cables, wireless communications networks and communications systems. If these satellite network capabilities were to be lost, then it would disable much of its response capabilities. The same is true for military systems all over the world.

The Future of Satellite Communications

Some forecasters have suggested that fiber optic networks with their terabits/second networking capabilities can and will replace satellite networking in the future. There are certainly applications where fiber optic networking will continue to predominate, such as in linking major urban centers and providing broadband networks in major cities. Satellites have a key future role to play in broadcasting, mobile services, and military communications, providing services in rural and remote areas, enterprise networks and multi-casting applications. There are many technologies that are keys to the future growth of satellite communications. These include digital compression and new digital multiplexing and coding systems, and new systems that allow operation in the progressively higher frequency bands, such as the Q, V, W and terahertz radio frequency bands as well as the optical bands.

The first applications for optical links were for intersatellite links, and this will likely soon be followed by the creation of a very high capacity optical ring that will affect extremely broadband information flows around the world. This optical ring was first envisioned as a military relay system known as the TSat, or Transformational Satellite System. Although this was canceled due to budget constraints, it is now being re-envisioned as a private commercial optical network with very similar engineering concepts. The so-called Halo optical satellite system would create a ring of ultra-high capacity networking around the globe as seen in Fig. 3.9 below. Even if this HALO network is not put in place, other optical rings in the sky will ultimately create a massive broadband cable in the sky.

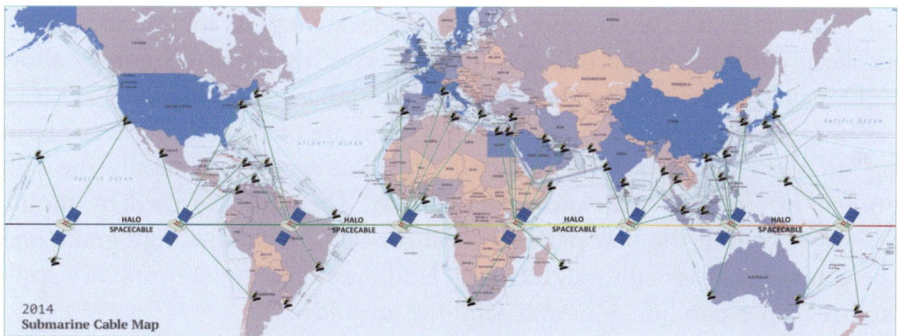

Fig. 3.9 The Proposed Laser Satellite Global Ring Architecture by Laser Light. The proposed Satellite Optical Space Cable in the Sky will have a worldwide reach. The main thing to note is that the move to higher and higher frequency bands will eventually give satellites the same throughput capability as fiber optical cables. Eventually satellite links will be needed to the Moon and Mars and even asteroids, where mining operations are taking place. This will be not just for exploration but to connect with habitats and mining colonies and off-world civilization. The off-world space economy will be important to the world, and satellite links and information networks will be a critical part of that future. The same is true for other existing commercial space applications and services (Image courtesy of Laser Light Communications.)

The Remote-Sensing Industry: The Second Commercial Satellite Market to Emerge

The first application satellites that were launched were communications satellites, but the next type of satellite that went up was created to observe Earth below. For many years people had been using kites, balloons, and airplanes to try to see the ground from above. This information was important to armies, to see their enemies; for farmers and foresters to see their crops and trees to spot possible disease; and for fishermen to locate schools of fish. The ability to see the ground synoptically has proven useful for just about everyone—urban planners, miners, oil companies, surveyors, military personnel, land speculators, and even retailers. Remote sensing is used today for many unusual purposes, such as the prosecution of war criminals, the study of climate change by scientists and plotting the spread of a pandemic by medical personnel.

Although the remote-sensing market is far smaller than the satellite communications market it, too, has shown consistent growth and is now a multibillion-dollar business. Perhaps more importantly remote sensing has become vital to more and more businesses around the world. McDonald's is one of the biggest users of remote sensing data. This fast food giant is not a big user in order to monitor potato crops, but to follow vectors of urban growth so to invest in land more efficiently. They see where growth is heading and then buy tracts of land that allows them to provide small plots for a new franchise and keep additional land to sell to developers. This space technology allows McDonald's to realize large profits as a real estate land speculator.

Commercial satellite remote-sensing systems are used today in "smart farming" to allow automated tractors and irrigation systems to supply the right amount of fertilizer and water to farmlands. They are also used by mining businesses to find the best places to drill for oil or seek new high grade ores. The commercial, agricultural, retailing, land development, and military applications keep expanding as new capabilities are added.

There are remote-sensing satellites that use infrared, optical and ultraviolent sensors. There are satellites equipped for active sensing using radar imaging that can penetrate cloud cover. There are even sensor systems that are optimized for the oceans and ice cap regions and some that can relay ground or ocean-based measurement up to the satellite from remote locations and then back to processing stations. One of the most important new capabilities is that demonstrated by Earth-observing satellites that can now perform hyper spectral sensing.

This newest type of remote-sensing satellites—rather than providing data across broad spectral ranges—are now gathering much more precise information

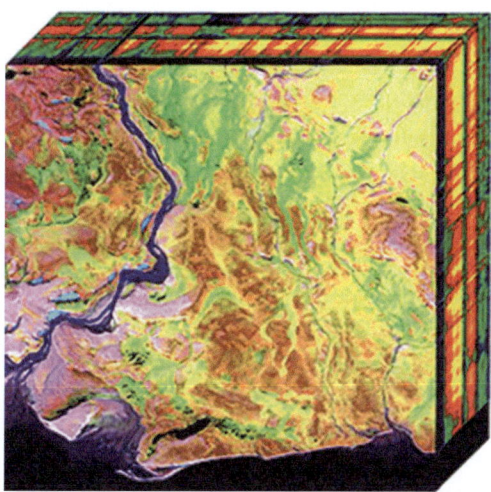

Fig. 3.10 The dense spectral data contained in a 3-D hyper spectral data cube (Image courtesy of NASA.)

across a wide spectrum of frequencies. These hyper spectral satellites break down their wide ranging optical sensing systems into 40, 50 or more very narrow and very specific spectral regions to produce much more finely tuned data. This type of sensing, after detailed processing and analysis, can tell where crops or trees might have a particular disease. They might reveal a particular type of pollutant in a stream. They can serve to provide remote-sensing analysts with precise information that comes from there knowing the "signature" associated with specific minerals, weeds, different types of grapes, corn, oats, and even fish or algae. This more precise remote sensing capability does require much more processing by analysts to derive all the detailed and subtle information these new sensors in the sky can produce (Fig.3.10).

Although the first remote-sensing satellites were projects of space agencies such as NASA, the European Space Agency (ESA), the Russian Space Agency (RosCosmos) and the Japanese Space Agency (JAXA), this activity has increasingly made the transition to commercial operations. Today commercial enterprises such as Spot Image of France, GeoEye and Skybox in the United States are offering remote-sensing services to commercial customers. Certain services that involve the construction of large and powerful remote sensing satellites such as RADARSAT 1 and 2 remain as governmental operations via the Canadian government. In Japan, China and India the operation of remote-sensing satellites and the processing of the satellite data is likewise a governmental operation. The longer term trends, however, seems to be toward commercializing remote-sensing operations, though conducted under strict governmental controls. The various commercial satellites operate under so-called "shutter controls" that limit the provision of imaging data in areas of armed conflict. In addition there have been restrictions as to the resolution—or level of detail—that can be used in imaging of commercial systems. These restrictions have in the past decade become less and less as technical capabilities in the satellites have increased. Some of the commercial systems can produce sharp images down to 25 cm per pixel, or about 1 ft per dot.

Remote-sensing satellites typically operate in low Earth polar orbits that allow sensors to remain "Sun synchronous or in geosynchronous orbits to provide a synoptic overview of wide regions of Earth. Meteorological satellites, which will be discussed a bit later, are essentially a special case of remote sensing and utilize these same type orbits. These meteorological satellites, rather than having a very broad focus, are today concerned with essentially three things only: Earth weather, and especially extreme weather conditions, solar weather (flares and coronal mass ejections), and climate change.

Governmental satellite systems that are deployed for military intelligence and strategic surveillance have even greater precision in their observational capabilities, with the ability to even read license plate numbers on vehicles.

The Satellite Navigation and Timing Industry

The precise satellite navigation and timing systems that operate today began with a particular military application in mind—namely the guidance and targeting of weapon systems. The initial U. S. Navstar system known widely as the Global Positioning Satellite (GPS) system was developed to allow precise targeting of missiles and aircraft bombers. At the time of the GPS system inception few people could have envisioned the remarkably wide range of applications that would eventually evolve from such a satellite network. The technology that allows them to operate are on-board atomic clocks that are precise to billionths of a second, and the propagation time for signals from a large constellation of these satellites allows some rapid computation to convert these times to travel from the satellite to the reception point to determine very exact locations. So today we use these systems to guide cars and trucks and buses to their locations more efficiently and soon will enable cars to move effectively around cities and countries without a driver at all. These systems are used for surveyors to determine boundary lines, for bankers and credit card companies to time stamp financial transactions, for pilots to take off and land aircraft more safely, even in bad weather, and by shipping fleets to avoid storms and come to port sooner.

There is a website devoted to explaining more than 100 applications for GPS and other precise navigation and timing satellite networks. The Soviet/Russian system known as GLONASS was the second such system, but today virtually every major spacefaring company has their own precise navigation and timing network. In addition to GPS and GLONASS there is the Galileo system in Europe that is just beginning to be deployed, the Japanese quasi-zenith system, the Chinese Beidou and Compass (i.e., Beidou-2) and the Indian Regional Navigational Satellite System. It is because these systems are considered vital to national security that so many of them are being deployed.

Geomatics, Geospatial Analysis and Geographical Information Systems (GIS) are three terms that one often now encounters when seeking to apply either satellite remote-sensing data or satellite navigation and positioning data both in a scientific or practical sense. The terms geospatial analysis and geomatics are applied to the integrated capabilities that can be created by effectively combining remote sensing, satellite navigation and other data by putting this information into databases. This is particularly so with regard to GIS, which is a type of information grid applied to the surface of Earth. This information processing format allows one to understand, interpret and use this data in a much more effective way. For many years there was the area of remote sensing or Earth observation that was entirely separate from precise navigation and timing satellite data, but the creation of the GIS database that allowed this data to be effectively combined has given rise to its integrated use into a single scientific area of study that is referred to either as geomatics or geospatial analysis. Each year the GIS process to organize data into geographic coordinates, first into two dimensions, and now into three dimensions has allowed satellite data to be used more and more effectively. Geomatics are now employed by police, fire and first responders, by farmers and miners, by military commands and strategic forces, by urban planners, by fishermen, and by hotels, restaurants and retailers seeking to announce their location to travelers. Many other scientific applications include zoology, biology, geodetics, meteorology, ocean research and climate change.

The added precision that is provided by the latest generations of precision navigation and timing (PNT) satellites, also known as Global Navigational Satellite Systems (GNSS), have allowed more and more applications. These vital applications include assistance with aircraft takeoffs and landings and guidance to driverless cars.

Today the launching of more and more PNT satellites has added to the precision of navigation and position determination, but has also led the United Nations to take the lead in seeking to coordinate the various national systems that are being deployed. This group, known as the International Committee on Global Navigation (ICG), was established in 2005 and participation is voluntary, but it also is used universally by all countries with GNSS networks. The vision statement for the ICG is as follows: "The International Committee on Global Navigation Satellite Systems (ICG) strives to encourage and facilitate compatibility, interoperability and transparency between all the satellite navigation systems, to promote and protect the use of their open service applications and thereby benefit the global community. Our vision is to ensure the best satellite based positioning, navigation and timing for peaceful uses for everybody, anywhere, any time."

The ICG meets once a year, and nations operating GNSS networks share information via a portal operated by the U. N. Office of Outer Space Affairs (OOSA). The biggest technical challenge is that the various systems operate in different frequency bands and in different orbits so that access by a universal GNSS receiver is quite challenging. There are many receivers that are dual use for the U. S. GPS and the Russian Glonass network, but many fewer that can access the other GNSS networks.

The U. S. GPS network was initially equipped with "selective availability" that could be turned on to provide much less accuracy for users without classified access. The United States, however, decided to not use this type of masking of the accuracy of their network during the Clinton administration. This was in part because of the many vital civilian applications that had migrated to the GPS network and in part because an engineer at the Jet Propulsion Lab had developed and patented a calculation system to defeat the "selective available" system.

Meteorological Satellites

The fourth major satellite application today involves meteorological, or weather, satellites. These satellites that are deployed in low Earth Sun synchronous polar orbiting satellites as well as in geosynchronous orbit provide a number of vital services that include weather monitoring and weather prediction, solar storm monitoring and tracking of longer term trends associated with climate change.

Again, most major spacefaring nations, such as the United States, Russia, Europe, Japan, China and India operate a network of meteorological satellites and also share data through the U. N. World Meteorological Organization (MEO) and the information sharing network of the WMO known as the World Weather Watch (WWW). The LEO polar orbiting satellites provide detailed and high resolution images of storm systems and have now special instrumentation such as lightning strike monitors that can track the movement of high intensity storms in near real time. The GEO satellites that provide an overall view of nearly one-third of Earth's surface at once provides the overall view of global weather patterns and assists with the integration of data provided from LEO satellites.

These satellites equipped with radiometers and other sensors can also detect changes to the Sun and provide warnings related to violent coronal mass ejections that can knock out power grids. Satellites that are not powered down during these powerful solar events are also quite vulnerable Further these meteorological satellites can monitor longer terms changes to weather and environmental conditions related to climate change.

Since extreme weather events are of mutual concern around the world and affect such things as airline and shipping safety, power grids and oil and gas line safety, there is systematic sharing of data concerning major storms such as hurricanes, typhoons and monsoons, as well as changes in La Nina and El Nino. In fact it is likely that the systematic sharing of meteorological satellite data is the most effective of all the application satellite operations and might be considered a model for the other types of satellite applications. In recent years the sharing of key data has improved to include not only data concerning Earth-based weather conditions but also solar storms and coronal mass ejections and climate change information as well. The U. S. National Ocean and Atmospheric Administration (NOAA), for instance, now operates a portal that provides near real-time reports on solar storm conditions.

Legal and Regulatory Concerns for Commercial Space Operations

The good news is that the various types of well-established satellite applications that are now vital to global safety and business operations have some 40–50 years of practical experience and that viable legal and regulatory processes exist to govern these services effectively. These processes can provide a useful path forward for New Space applications that are just evolving, such as solar power satellites, space mining, and space habitats and private off-world enterprises.

Areas that will need attention in future years include:

a. **Better and more authoritative international means to allocate electromagnetic (EM) spectrum.** This means better ways to meet the needs of current and future satellite applications so as to promote future growth and innovation, while also protecting the common interests of humankind.

(This means allowing future growth and more efficient use of the spectrum for space communications, remote sensing, space navigation, power transmission, and other needs in an equitable and legally enforceable manner.)
b. **Eliminate or reduce EM interference.** This means finding technical, political, financial or legal means to stop space systems from interfering or jamming one another or interfering with ground-based systems.
c. **Finding ways to make sure that current or future space industries play nice with one another.** The key here is to find a way to avoid egregious harm between New Space industries, established space industries, and Earth-bound enterprises. In the long run one might hope for a legally binding and economically efficient way forward, but this will take time. True space law that is effective and definitive would imply moving beyond the somewhat chaotic environment that exists today. It would imply the ultimate establishment of a recognized legal body that can impose fines or sanctions, or reallocate resources. Such a body would need to have the authority to avoid conflicting usage or exploitation of resources needed either on Planet Earth or in space. Areas of conflict to be resolved here could involve radio frequencies, light spectrum, sources of solar or nuclear energy, or access to or usage of off-world habitats, celestial bodies or resources.

In many ways the practical use of outer space is similar to exploiting the oceans for commercial purposes. In the case of oceans there are legal processes, however, that are enforceable, but in space there are really no such definitive bodies that can decide right and wrong and impose penalties. Until there are "space courts," with enforcement powers and the ability to provide sanctions and fines, the realm of outer space will be quite hard to "govern." The alternative will be to largely fall back to enforcement through national laws, or "codes of conduct" that spacefaring nations agree can be universally applied. We still have a long ways to go.
d. **Keeping pace with rapidly evolving technology.** The biggest challenge of all may be keeping pace with new technology. Today new, extremely broadband high throughput satellites and new mega constellations in low Earth orbit optimized to provide Internet services can present a range of technical, standards and regulatory challenges and issues. Orbital debris, orbital crowding, power limits and interference are just some of the concerns. New high-powered millimeter wave and terahertz satellites and new optical laser ring satellites may present new concerns as well.

Final Conclusions and Observations

In the absence of other ways forwards, national space laws—"model laws"—widely agreed codes of conduct, and agreements, standards or processes will create quite a hit-or-miss process about what to do about important space applications. The hard part will be to decide what to do when conflicts arise. The five U. N. treaties on space provide no more than a few rules and guidelines about the most basic issues involving the "enforcement" of what might be called space law. The truth is that space law, when compared to a national legal system, has numerous cracks and holes in it. It is like a house that has elements of a basic structure in place, but no roof, no flooring, and no bricks and mortar to constitute the walls.

The near-term prospects of a new, more detailed and definitive set of enforceable space rules seem at this juncture of U. S. history very far from an achievable goal. In the absence of a new and definitive set of space laws, one will have to make do with what is around today.

This means reliance on such organizations and processes as offered by the International Telecommunication Union (ITU), the Internet Engineering Task Force (IETF), the International Standards Organization (ISO), the International Electro-Technical Commission (IEC), the Institute of Electrical and Electronics Engineers (IEEE), the European Telecommunications Standards Institute (ETSI), the International Committee on Global Navigation Satellite Systems (ICG), the InterAgency space Debris Coordinating (IADC), and the U. N. Committee on the Peaceful Uses of Outer Space (COPUOS). For better or worse we will need to rely on a patchwork of standards and "quasi-agreements" that allows some modicum of agreement. Satellite communications, remote sensing, satellite navigation, and meteorological satellite services have somehow managed to be launched and operated with some degree of effectiveness for over 40 years, and commercial enterprises—especially in the satellite communications business arena—have not only emerged but thrived without definitive space laws. Other new space businesses may well be forced to do the same. These new space enterprises will be able to build on the foundations that these early space industries have painstakingly built decade after decade. Most businesses, whether in space enterprises or not, very much desire legislative and regulatory certainty and widely accepted standards.

Nevertheless, there is a process to define which frequencies can be used and for which application. Further, the U. N. space treaties from the 1960s and 1970s provide some reasonable basis to move forward, but as noted in Chapter Nine, these international agreements are beginning to show their age. The International Telecommunication Union currently lacks enforcement powers, but it has largely coped with radio frequency spectrum allocations, developed standards for the operation and minimization of frequency interference for various types of satellites, and helped to bring a set of multibillion space industries to life. It is not perfection, but it is a start.

References

1. Issie Lapowsky, "The Start Up That Could Beat SpaceX to Building a Second Internet in Space," *Wired,* January 22, 2015. http://www.wired.com/2015/01/greg-wyler-oneweb/. Last accessed May 21, 2016.
2. James O'Toole, "Google buys satellite start-up Skybox Imaging" *CNN Money,* June 10, 2014. http://money.cnn.com/2014/06/10/technology/innovation/google-skybox/index.html. Last accessed May 21, 2014.

4

Commercial Space Transport, On-Orbit Servicing and Manufacturing

Introduction

The key springboard to the development of new commercial space activities are improved and less costly space transportation capabilities. Zooming into space for a fraction of today's cost is clearly the way to the future. You can't create a New Space economy if you don't have space vehicles that are safe, reliable and cheap.

Fortunately a New Space transportation revolution has already started. Today there are a variety of new launcher systems ready to make either low cost orbital or suborbital flights. Even to those in the space business, the names that are out there today were unfamiliar in the year 2000. The New Space transportation revolutionaries include SpaceX, Virgin Galactic and Launcher One, Sierra Nevada, XCOR, Rocket Lab, S-3, Reaction Engines, Blue Origin and quite a few more. These radical new companies are pioneering low cost ways to get to low Earth orbit via low cost launcher systems or developing suborbital spaceplanes. The spaceplanes, such as Space Ship 2, are currently designed for so-called space tourism, or more accurately, "space adventure" flights. But this is just the start of a multi-billion dollar enterprise. There are now focused efforts to develop hypersonic transport systems to allow 3-h parabolic flights through protospace (or near space) from London to Sydney.

Feed into the mix those that are creating spaceports, private space habitats and new launcher systems to GEO orbit and beyond and you get a sense that new commercial space is shaking up the status quo. Just the latest headlines are that Elon Musk and his company SpaceX are planning to send his Falcon 9 Heavy and a Red Dragon capsule to Mars. Not to be outdone billionaire Robert Bigelow, head of Budget Suites and Bigelow Aerospace, is intent on creating private space habitats with more internal volume than the International Space Station. Currently his inflatable habitat system is being tested on board the International Space Station.

Then there is the world's largest aircraft and high altitude rocket launcher that is nearing completion under funding provided by billionaire Paul Allen and his Vulcan Enterprises aerospace company. More about this to come.

Even NASA has jumped onto the New Space ventures bandwagon and offered competitions with prize money to develop new launcher technology. They have also awarded massive development contracts to SpaceX and Boeing to build new commercial launch systems to get astronauts to the International Space Station and back. When new entry Sierra Nevada formally protested that their Dreamchaser spaceplane should have been one of the finalists, NASA reconsidered and has provided them with a hefty development contract as well.

And others are getting into the act. Alan Bond's Reaction Engines Ltd. is developing the single stage to orbit Skylon spaceplane in the United Kingdom. Bristol Aerospace Ltd., under European Space Agency funding, conceived of a 300-person hypersonic spaceplane to provide new hypersonic transport capabilities. Space Swiss Systems (S3) is developing spaceplanes for space adventures, but with an additional stage that will enable it to launch small satellites into low Earth orbit. And these are just some of the New Space enterprises that span the globe.

The bottom line here is that the world of space has fundamentally changed. The three decades marked by space shuttle operations from the 1980s through the early 2010s were interesting and exciting, but the period since the shuttle was grounded in 2012 has produced a plethora of innovation in commercial space transportation and other exciting New Space ventures. This new commercial space revolution bespeaks a totally new era of space travel and the rapid coming of age of new low-cost space travel.

And the innovation does not stop with just the invention of better rockets. There are scientists and engineers that are exploring even more exotic ways to reach into space using tether technology and a combination of lighter-than-air craft and ion engines in order to get to low Earth orbit. Others are seeking reliable and cost effective ways to repair or upgrade satellites in orbit. Yet other space engineers are exploring ways to move satellites in failed orbits to their proper location, or even to remove dangerous derelict satellites from space before they hit another satellite or upper stage rocket and create a dangerous new shower of space debris. Some are even thinking about ways to build new "space shields" to ward off asteroids and solar storms. The bottom line is that "space futures" are simply not what they used to be.

4 Commercial Space Transport, On-Orbit Servicing and Manufacturing

It is a whole new world in commercial space transportation, and the innovation just keeps bubbling forth. These amazing new inventions in space transport, in space habitats and on-orbit servicing are truly setting the stage for a host of New Space ventures. Every day there are New Space startups spinning off from institutions such as the International Space University, the Singularity University, the Global Space Institute and places such as Silicon Valley and the Ames Space Research Center in Mountain View, California. The menu of possible avenues to fly off Earth's surface for business and pleasure is longer and more exciting than ever before. The Google Lunar XPrize alone has spawned more ideas in space transportation than the world could ever imagine. This chapter explains where we are today and where we are going.

The main topic of this chapter is new commercial space transportation systems, but we will also conclude by addressing with what may become possible with on-orbit robotics, 3D printers and fabricators. With a combination of low-cost space transportation and new robotic technology, we will be able to undertake on-orbit servicing, upgrades and repositioning of satellites and even have the possibility of using 3D printers in space to create replacement parts for broken satellites with failed components. Remotely controlled factories in the skies may ultimately be a part of the new gold rush.

The New Players in Space Transportation

As we have already explained New Space transportation capabilities are sprouting up everywhere. Space Exploration Technologies Corporation (SpaceX), Virgin Galactic, and now Blue Origin are all headed by space billionaires. These New Space titans draw a flock of news reporters like honey attracts a greedy Pooh Bear. Their exploits, however, generate ink for headlines for good reason. Elon Musk, head of SpaceX; Sir Richard Branson, head of Virgin Galatic; and Jeff Bezos, head of Amazon.com and the innovative Blue Origin rocket company in West Texas are among the top leaders re-inventing the space industry. Yet this trio of billionaire space zealots are just a few of the entrepreneurs and New Space ventures that are changing the face of what might be called the new cosmic enterprise.

Swiss Space Systems (S-3), Rocket Lab, Reaction Engines, Ltd., Sierra Nevada Corporation, XCOR, Orbital ATK, and many others are today intent on revolutionizing our ideas about the New Space industry and what it means for the future. It is not widely recognized, but Google, Facebook, and other Silicon Valley technology companies have awoken to the potential of space systems as well. They, too, are now in the space business or on the verge getting into it.

Yet another space billionaire, Paul Allen, the co-founder of Microsoft, who bankrolled the first spaceplane, "Space One," back at the start of the millennium, is now working to build Stratolaunch. This is a high-altitude aircraft designed as rocket carrier and launcher that can make large rocket launchers more cost effective. This is similar to Burt Rutan's White Knight carrier that carried SpaceShipOne aloft but very large.

Allen's exploits just add to the mix of the other corporate titans who are propelling new commercial space ventures to create a New Space economy (Fig. 4.1).

And even well-established aerospace companies such as Boeing, Lockheed Martin, Air Bus (EADS), Northrop Grumman and Orbital ATK are intent on re-inventing themselves to be able to produce New Space technology that is better, faster and lower in cost. Currently SpaceX and Boeing are competing head to head to produce proven crew-rated launcher systems to ferry astronauts to and from the International Space Station (ISS) beginning in just 2017 or 2018.

Altogether there are over two dozen new commercial space transportation companies around the world vying to provide new commercial launch capabilities. Going through a complete laundry list of all the contenders and wannabes is really not helpful in explaining the take-off of the new com-

Fig. 4.1 The Stratolaunch Mega Jet designed to assist airborne rocket launches (Image courtesy of Stratolaunch.)

mercial space industry. Likewise listing all the members of the Commercial Spaceflight Federation is not as important as noting that there are over 30 full or associate members to the CSF, and the numbers keep growing.

These details are not important unless you are seeking to be one of the new contenders yourself. For those seeking details about all the new commercial space entries and where all the various spaceports are being licensed in the United States and around the world, get a copy of *Launching into Commercial Space* by Joseph Pelton and Peter Marshall and the websites listed in the second edition of this book.

The Challenge of New Space Transportation Systems

The key point to be noted here is that there are dozens and dozens of these New Space organizations, and the numbers keep growing. Although they are largely U. S. based, there are New Space projects around the world. In some countries, such as Japan, India, China and Russia, most of these efforts are still managed and largely controlled by governments, but in the United States, Canada, and Europe private enterprise is largely leading the way.

And the pathway forward has not always been easy. There were two back-to-back key failures with new commercial space vehicles. First there was the Oct. 28, 2014, launch failure of the Orbital Sciences (now Orbital ATK) Antares vehicle that was to carry a resupply mission to the International Space Station [1]. Three days later there was the Oct. 31, 2014, crash of the Virgin Galactic SpaceShipTwo, due to pilot error. Rick N. Tumlinson, Chairman of Deep Space Industries, said in response to the SpaceShipTwo crash he was relieved about the initial response in the media to the accident in not overstating its importance. He and other New Space colleagues have emphasized that the commercial space industry will need to work harder to demonstrate its commitment to safety. "The thing to do is to acknowledge the challenges we're facing and make sure we're doing everything possible to mitigate those challenges using the best technologies, the best systems, and the best approaches… Let's not make the same mistakes that led to things like Challenger." Tumlinson added that this accident may provide a stronger argument for calling private human spaceflight something other than "space tourism." That term, widely used both within the industry and among the general public, makes the experience sound safer than it is likely to be for years to come [2].

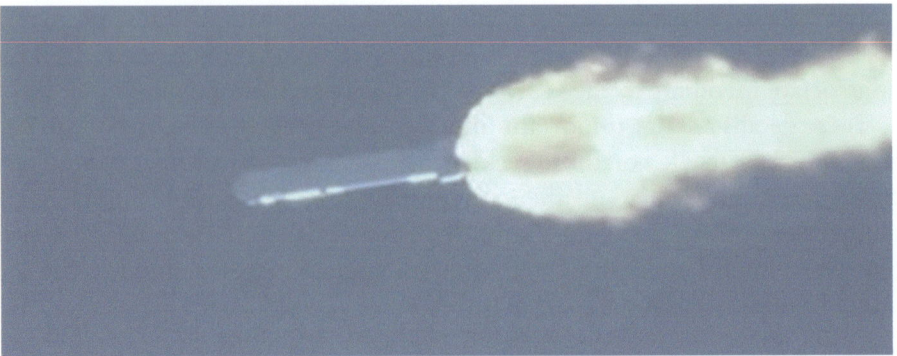

Fig. 4.2 The spectacular crash of the Falcon 9 after 18 successful previous launches (Image courtesy of SpaceX.)

Then in July 2015, after 18 previous successful launches of the SpaceX Falcon 9 launch vehicle, this new commercially designed launcher also failed (see Fig. 4.2).

These three failures by commercial New Space launch systems (the SpaceShipTwo spaceplane, the Antares launcher and the Falcon 9) clearly demonstrate that commercial launchers are not infallible, although hope remains that new technology, new streamlined designs and new management systems can produce launchers that are safer, more reliable and lower in cost than those produced by the space agencies. The space agencies have always placed great reliance on very high component and manufacturing standards and quality assurance processes. The commercial sector has, in contrast, sought simpler and newer designs. It seems that there is no magic answer, and that there is much to be learned from the techniques and approaches used by both space agencies and the New Space industries. The good news is that the failure analysis for the Falcon 9 has produced a clear understanding of the strut failure that ruptured the fuel tank that led to the failure, and this system will be redesigned to avoid similar future failures [3].

Since the crash SpaceX has developed the Falcon 9 Heavy vehicle, which is equivalent to the largest of the U. S. rocket launchers. Most recently Musk has announced plans to launch in 2018 this largest vehicle to Mars with a "Red Dragon" capsule on board. The objective is to demonstrate the ability to eventually send astronauts to Mars. As Musk has said with serious intent but tongue in cheek: "I want to die on Mars, but not on landing." This remarkably rapid progress by the private space industry is impressive, but it reopens the tough to answer question: "Why cannot NASA make such remarkable progress at a similar rate and such low cost efficiency?" In fact the latest question

4 Commercial Space Transport, On-Orbit Servicing and Manufacturing

that surfaced is, why does NASA need to spend billions to design and build the new Space Launcher System (SLS) if SpaceX's new heavy lift launcher can go to Mars? [4]

The exploits of Elon Musk have frequently dominated headlines, but he is far from alone in his quest to conquer space and use his entrepreneurial skills to produce impressive space exploits. The New Space ventures range from very large companies with tens of thousands of employees and a long track record in aerospace at one end of the spectrum down to small start-ups that have a dream of creating the perfect new technology that makes future launchers better, safer and more cost efficient.

Questions still abound. What is the scope and objective of the New Space ventures? What is the ultimate end objective? Is the idea just to find a new way to make money and create new jobs? Or is the goal simply to create spaceplanes that give space tourists a 4-min thrill of weightlessness at the top of a parabola some 100 km up in space? Perhaps the mission is to create hypersonic jets that can span oceans in a few hours, or overcome the perils of climate change and effect a planetary defense against asteroids and violent solar storms. Perhaps the long-term goal is to allow humans to go into space forever and create colonies on the Moon and Mars and truly to explore the cosmos? Or is the idea truly to do all the above and more? Is the ultimate objective to realize the concept of Stephen Hawking to allow human civilization to spread the genes of Homo sapiens beyond Earth in order to allow the human species to survive for eons and eons to come by going beyond the Solar System to sow human seeds across the Milky Way Galaxy and beyond?

In short, what does it really mean to go into outer space? And is the New Space revolution truly going to change human destiny forever? Certainly questions abound.

If fat cat passengers with the wherewithal to pay $250,000 can fly above 100 km and return, do they become "citizen astronauts" with some special status when they come back? Will we soon have safe aircraft that rise to the edge of outer space at Mach 6 that might allow passengers to fly across oceans in a few hours and perhaps become "citizen astronauts" on their travels that just brush outer space?

If we look a bit further into the future, will we ultimately see commercial carriers offering flights to space stations or the Moon as first envisioned by Arthur C. Clarke in *2001: A Space Odyssey* and then repeated in science fiction movies such as *The Sixth Element*? Is space tourism just to experience weightlessness, experience the northern lights and see Earth against the dark sky of space in an experience that lasts only a few minutes? Or can we evolve toward something more profound? Or could we see, within a generation or so, flights to off-world locations for legitimate and commercially valid reasons associated with space mining, space manufacturing or commercial experiments carried out on private space habitats?

The list of questions about New Space and the new thrust to create new commercial space transportation systems is almost endless, but most of these questions start with the fundamental need of viable, safe and low-cost space transportation. If this cannot be achieved, then none of the above visions can ever be realized. Reliable and significantly lower cost launcher systems represent the prime gateway into a New Space world.

Beyond these developments then, grander visions such as "space elevators" and space tether lift systems may also be possible in the longer term. Space habitats, space mining, solar power satellites, and more start with the right space transport. Let's explore progressively where we are today, starting with suborbital spaceplanes and launch systems. Next we will examine new commercial capabilities for launching into low Earth orbit and finally get us easily to the Moon, to reach near Earth asteroids (NEAs) and Mars and go even beyond.

Spaceplanes and the Space Tourism Business— The Start of a New Space Adventure

Arthur C. Clarke's "Three Laws of Prediction," although witty and tongue in cheek, are also amazingly insightful. His first law is a particular case in point: "When a distinguished but elderly scientist states that something is possible, he is almost certainly right. When he states that something is impossible, he is very probably wrong."

The wisdom of Clarke's First Law particularly hits home when it comes to the topic of space tourism. Some 18 years ago, when Peter Diamandis, the father of the XPrize, came to talk to the Masters' students at the International Space University (ISU) in Strasbourg, France, while this author was serving as dean, he made a pitch for their joint class project to be a study of space tourism. The students were instantly enthused and were all in favor of this topic. But the "senior faculty" intervened and said, no, this topic is too much like science fiction and will never be accepted as a serious academic work. The students were forced to design a better global satellite navigation system instead. Today, looking back, the students and space visionary Peter Diamandis were right, while the senior faculty were much too cautious.

Peter Diamandis went on, just 5 years later, to convince the Ansari family to finance the $10 million XPrize. This initiative had first been announced among much fanfare in 1998 at the Smithsonian National Air & Space

4 Commercial Space Transport, On-Orbit Servicing and Manufacturing

Museum and was fully funded by 2001. This was just a modest few years after the students at the ISU had been told spaceplanes and space tourism was too much of a science fiction dream to be considered a class project that would assess the feasibility of it from a technical, operational, financial and safety perspective.

Over a dozen organizations proceeded to sign up to try to compete for the XPrize that, again, most of the "wise experts" thought was impossible to win because a winning spaceplane had to be designed, built and successfully flown within just 3 years. Paul Allen, co-founder of Microsoft, however, financed Burt Rutan and his team at Scaled Composites to try. The Rutan team designed the SpaceShipOne spaceplane system that flew into space to an altitude above 100 km and then flew again with a crew within a 5-day period to meet the contest requirements. When the second flight landed on October 4, 2004, an impromptu sign was held up within the crowd that had assembled in the Mojave Desert: "SpaceShipOne, NASA None" (see Fig. 4.3).

From the start, the SpaceShipOne flight and the Ansari XPrize was based on an unconventional start-up entrepreneurial model. Not only were all the teams that competed for the prize composed of private aerospace designers and engineers, there was no participation by official governmental space agencies.

Then there is the story of Texas-based Ansari family, which was actively courted by Peter Diamandis to finance the project. The Ansari family came up with a clever ploy to establish the prize for which the teams were to compete.

Fig. 4.3 SpaceShipOne on display at the Smithsonian National Air & Space Museum after its historic flight on October 4, 2004 (Image courtesy of the National Air and Space Museum.)

They did not put up $10 million. Instead they bought an insurance policy for about $1 million from an insurance group that was confidently assured by the "wise experts" they would never have to pay the $10 million prize money, so it was actually easy money. But in the end it was the insurance company and not the Ansari family that had to fork over the prize money. Since Paul Allen probably paid Burt Rutan at least $30 million to $40 million the entire exercise ended up not being about money or profit, but about re-inventing ideas about how spaceplanes could be built and flown.

Since the start of the twenty-first century, a whole new spaceplane and space tourism industry has grown up out of nowhere. It turns out that there were a lot of would-be space entrepreneurs who had unfulfilled aspirations. These are the outside-the-box thinkers who see their mission as being to re-invent the future.

In addition to Peter Diamandis, one of the other pioneers was Eric Anderson. Anderson left NASA and launched a company called Space Adventures. He decided to approach Roscosmos, the Russian space agency, to book private astronauts to fly on Soyuz launch vehicles for a stay on the International Space Station. These "citizen astronauts" were somehow persuaded to pay amounts that started at $20 million a trip, which escalated to $25 million, $35 million, $40 million and then over $50 million. Eric's company is even offering a trip around the Moon for over $100 million.

At the start Eric went to scientists and financial fund manager Dennis Tito to get funding to launch his business. Tito told Eric: "I really don't want to finance your company, but I will pay to fly into space." Eric negotiated with the Russians and came back and said that for $20 million he could fly up and stay on the International Space Station. Dennis Tito agreed to the deal. This was perhaps the second great boost to New Space activities comparable in importance to the XPrize competition.

Thus Space Adventures was launched, and the idea of "space tourism" went from dream to reality. Since then Mark Shuttleworth of South Africa and Greg Olsen, a high quality camera manufacturer from California, flew, and then the fourth private astronaut was a woman, none other than Anousheh Ansari, who along with her brother, had bought the $1 million insurance policy to fund the Ansari XPrize. The first four flights reportedly went for $20 million, but then inflation set in big time. Number five and number six were both for Charles Simonyi, Ph.D., who developed some of the earliest Microsoft software and notably was dating TV celebrity Martha Stewart during his flights. Purportedly Martha prepared some special duck pate for his flights. Simonyi paid $25 million for flight number 5 and then ponied up $35 million more for flight number 6. Two single seats for these two flights to orbit came in at $60 million.

Gary Gariott, the son of astronaut Owen Gariott, went up in 2008 for $35 million as well. Then Cirque du Soleil CEO Guy Laliberté paid $40 million for his trip to space in 2009. But since then there has been a lull. Reportedly the price tag for the scheduled September 2015 flight by opera singer Sarah Brightman, star of *Phantom of the Opera,* had risen to a spectacular $52 million. But then she unexpectedly canceled her flight in May 2015. She had worked with Andrew Lloyd Webber to sing a special song in outer space but then reversed plans for "personal reasons." What refund she might have received is not known.

Part of the reason why there may be fewer people willing to pay a king's ransom of over $50 million to fly to the space station could be the pending Virgin Galactic flights at much lower fares. Apparently notables such as Tom Hanks, Justin Bieber, and Victoria Principle have signed up with Branson's company for a suborbital flight to swim in space for 4 min and see the big blue marble against the dark sky of space for a ticket price of $200,000 that has now escalated to $250,000. It is noteworthy that well over 500 people have booked for flights on Virgin Galactic and that the number of the bookings well exceeds the total number of all astronauts from all countries around the world that have ever flown to date. Within the decade the whole significance of being someone that has flown in space could be hugely depreciated. One might 1 day hear the following conversation: "Are you going skiing in the Rockies this winter?" "No I thought I would take a flight into outer space to see the Aurora Borealis, instead."

Peter Diamandis and Eric Anderson have worked in tandem to spur the space tourism industry. They were strong advocates to launch what was first called the Private Spaceflight Federation. This then quickly evolved to become the Commercial Spaceflight Federation that now includes a collection of spaceplane developers, spaceport operators and a variety of aerospace organizations. XCOR, with a variation on the concept of taking a spaceplane ride into "outer space" is offering a flight on their Lynx Mark 11 that only goes up about 40 miles (64 km) for only some $100,000.

For those on a limited budget but a zeal for outer space there are options. Peter Diamandis' company, known a Zero Gravity, flies passengers that wish to "fly in space" on his so-called "vomit comet." This type of flight involves a series of parabolic flights. On each parabola one gets about 20–40 s of weightlessness on each of up to 18 arcs. These flights can be booked through Space Adventures for only about $4000 to $5000. The most common arrangement is for corporations or organizations to book a flight for employees (see Fig. 4.4).

Yet another alternative to flying on Virgin Galactic's SpaceShipTwo is to sign up to fly in a Russian Foxbat jet up into the stratosphere where one can see the curvature of Earth and dark sky for a cost of around $10,000. XCOR

Fig. 4.4 Zero G weightlessness flights represent a "cheap option" for those who cannot pay a quarter of a million dollars to fly with Virgin Galactic (Image courtesy of Zero Gravity Corporation.)

will soon provide a low altitude flight to half the altitude for about half the money. Actually a range of experiences are available via Space Adventures that also include cosmonaut training in the Russian Star City near Moscow. In talking to Peter Diamandis, he confirmed that the thrill of flying with Stephen Hawking on one of his Zero G flights was a highpoint of his career—both figuratively and literally.

The bottom line is that there are many, many efforts around the world to develop what is often called the space tourism experience. These efforts include zero g flights, flights in high altitude jets, astronaut training, and plans to create new spaceports that are sort of a Disneyland for space enthusiasts. In parallel there are serious efforts by entrepreneurs and space and aviation agencies to develop spaceplanes and hypersonic aircraft that could provide flights across oceans.

Japan's space agency, JAXA, as well as the Germany's ESA-DLR Space Institute, with the TALIS Institute as well as NASA, are all now into testing spaceplane prototypes that can fly in the Mach 2 to Mach 4 speed range. Part of these space agencies' missions are to find spaceplane systems that are safe and non-polluting, while other research is devoted to creating "nose needles" that can be extended from the craft to create a series of minor sonic booms rather than one massive sonic boom during landings. One reason for stopping Concorde flights to the United States were the very loud sonic booms that created havoc when the Concorde took off and landed. The mini-booms are thought to present a solution to this problem.

Fig. 4.5 Artist's rendering of a JAXA prototype of H_2-O_2 fueled hypersonic jet (Image courtesy of JAXA.)

There are an even larger series of efforts to develop commercial spaceplane craft. The Japanese agency JAXA provided a detailed briefing at the ICAO meeting on space traffic held in Montreal, Canada, in 2015. They spoke of their rapid progress in developing high-altitude hypersonic jets fueled by liquid hydrogen and oxygen, designed for intercontinental transportation (see Fig. 4.5).

Just some of the newest projects include S-3 in Switzerland, External Engines, Ltd. and Bristol Spaceplanes in the United Kingdom, Air Bus-Astrium, QSST, and Lockheed. These development programs are in addition to XCOR and Virgin Galactic and their spaceplane development. Many of these projects are targeted at transportation needs rather than space tourism flights. This is simply because this is where the money is. The schedules for actual test flights with passengers are still far from certain, but certainly several years in the future.

There is one system under development that does not fit the spaceplane model. This is the development by Jeff Bezos' Blue Origin company of their New Shepherd reusable Vertical Takeoff and Vertical Landing launcher system. This unique space tourism system involves both a reusable rocket launcher and capsule that is launched and separated from the New Shepherd to allow paying passengers to gently parachute back to Earth while enjoying their outer space experience through large observation windows. The demonstration of this entire sequence by Blue Origin in November 2015 puts this new type of launcher and capsule system into serious contention as well.

New Low Cost Launchers to Earth Orbit

The physics, velocities and thrust components for an airplane, a conventional jet, a spaceplane, and rocket launcher into Earth orbit are significantly different. The specific energy needed to attain orbit rather than just undertake a suborbital flight are on the order of 25 times greater. Spaceplanes represent a significant accomplishment and could be used for hypersonic flight and other important tasks, but launching satellites or crewed missions into orbit require five times more velocity and a great deal more thrust. Table 4.1 below shows the major quantitative differences represented by airplanes, jets, spaceplanes and rockets.

Nevertheless several of the developers of spaceplane systems have in mind to extend the capabilities they have developed for suborbital flights to create at least the ability to launch small satellites. Virgin Galactic has entered into contracts with OneWeb, the mega-LEO communications satellite constellation, to use its Launcher One system to boost 125-kg small satellites into orbit in the 2018 time frame. The Swiss Space Systems, or S-3, organization is seeking not only to engage in space tourism suborbital flights, but also add a capability to launch small satellites.

What seems even more ambitious are the efforts of companies such as Firefly and Space Lab, which are seeking to create entirely new launch vehicles for what is definitely the low end of the market.

NASA in October 2015 awarded a total of $17.1 million in contracts to three companies for small satellite cubesat launches. One contract for $6.9 million went to Rocket Lab USA, which is based in Los Angeles and is developing the Electron rocket launcher that will launch from New Zealand. This launcher, with its so-called Rutherford "electric motors," is the first of its kind. Another contract, valued at $5.5 million, went to Firefly Space Systems of Cedar Park, Texas. The final contract of $4.7 million went to Virgin Galactic LLC of Long Beach, California, for Launcher One launches. These NASA contracts were awarded under fixed-price "Venture Class Launch Services" (VCLS) contracts.

However, most public and media attention is not focused on cubesat launches but on rockets that can lift humans into space. The question that remains to

Table 4.1 Comparing airplanes, jets, suborbital spaceplanes and rockets to orbit[a]

Comparative Factor	Airplane	Jet	Spaceplane	Rocket to low Earth orbit
Velocity (m/s)	250	500	1600	7800
Height (km)	Up to 10	Up to 20	Up to 120	200+
Specific energy (J/kg)	0.13	0.7	14.5	324

[a]Data provided by Professor Nikolai Tolyarenko of the International Space University

the entrepreneurs is, where is the market that can support these new ventures? Is it space tourism, is it small satellite launches, or is it hypersonic transport? Currently the developers are trying to straddle all three possibilities.

In addition to Virgin Galactic, which has the lead in the space adventures market, there could be others that are developing spaceplanes, including and XCOR. Then there is Skylon, which seeks to fly into orbit with a single-stage scramjet system, and yet others such as Blue Origin, which is trying to develop the New Shepherd and Sierra Nevada with its Dream Chaser, which might be able to support all three markets. To date only Sierra Nevada has indicated that it could create a capability to fly a human crew to low Earth orbit.

Chuck Lauer of Rocketplane was recently at an International Association for the Advancement of Space Safety (IAASS) conference. He indicated that his company was focusing on small satellite launch capability because this was a clearly emerging market to be serviced and the liability and safety issues were far less demanding.

New Space Entities with Major Launch Capabilities

The most significant launcher development in terms of larger scale space missions by any New Space entity is undoubtedly that of SpaceX. In the past decade SpaceX has developed in rapid succession: the Falcon I, the Falcon 9, and now the Falcon 9 Heavy. Elon Musk and his team in California and Texas have accomplished what no one thought possible. This is the creation of a reliable and low-cost new launcher system that can not only compete with established U. S. aerospace companies such as United Launch Alliance, Orbital ATK, etc., but actually compete in the world launcher market and provide launch services at the most competitive price. The Falcon launchers can thus compete with Chinese, Indian, Japanese, Russian and Ukrainian rocket systems. The failure of the Falcon 9 on June 28, 2015, due to the breakage of a support strut that secured a helium tank clearly has set back the SpaceX launcher development program and adversely affected the schedule for resupply of the International Space Station, plus had a key impact in the delayed deployment of the Iridium NEXT mobile satellite communications systems. Even so there is no doubt that SpaceX is now seen as a world-class supplier of launch services.

The latest innovation that has come from SpaceX is its ability to re-land its Falcon 9 launcher both on the ground and on a sea-based platform. This is seen as the first step to providing reusable rocket launch services. SpaceX has indeed indicated that the process of developing reusable vehicles could reduce cost by on the order of 30 %. It stated that its objective is to get launch costs down to $2000 per kg, or $1000 per pound.

When Elon Musk speaks about much more cost effective launchers in addition to high profile missions to Mars others people listen. New start-ups such as Rocket Labs USA, Firefly, Blue Origin's New Shepard, Space Swiss Systems, and even Virgin Galactic's Launcher One also all hold out the promise of lower cost and reliable space launch capacity. They, too, must be taken seriously. Clearly launch costs are headed south.

In short, innovation in launch systems now seems to be everywhere. The Rocket Lab design of the so-called Rutherford engine seems particularly interesting. The Electron vehicle utilizes an electric turbo-pumped LOX/RP-1 engine specifically designed to produce a high level of specific thrust and do so for a sustained period of 4 min. This involves an entirely new propulsion cycle that uses brushless direct current motors and high performance lithium polymer batteries that drive the turbo-pumps. This seems to be a more efficient and less polluting launcher design. The old model, where space agencies worked with huge aerospace companies to build rockets and high tech satellites, seems to be fading from the scene. Instead new entrepreneurial companies that are small, innovative, willing to embrace totally new technology and re-invent systems with totally new architectures and designs are now being considered for contracts along with the big guys. The approach might be called "the Silicon Valley model," and the attitude can at times be brash, arrogant—and successful.

Brett Alexander, at the time the President of the Commercial Spaceflight Federation, put things quite bluntly. He said: "The space agencies have their way of doing things and we have ours. The space agencies have managed to kill about 4 % of their astronauts on space missions to date. We want to invent New Space technology that is safe enough for everyone to use and doesn't cost billions."

It is a sign of the times that NASA is embracing the new guys on the block and funding open competitions and awarding these new "Venture Class Launch Services" contracts to totally new companies from the New Space world that are not afraid to reinvent launch systems technology.

The New Ventures in Protospace

And the New Space venture opportunities will not all just be in the areas we commonly refer to as "outer space." There are new efforts afoot in what some call the "protozone," or "near space." Recently the French government

4 Commercial Space Transport, On-Orbit Servicing and Manufacturing

Fig. 4.6 Artist's depiction of Stratobus in operation (Image courtesy of Thales Alenia.)

contracted with Thales Alenia to develop and test what Thales has branded the "Stratobus." This High Altitude Platform System (HAPS) is designed to operate in a constantly maintained stable location at about 20-km altitude. At this location winds are fairly stable and seldom exceed 90 km/h. This altitude, with a 500-km field of view, allows communications or observation over an area of about 180,000 km^2 (72,000 sq. miles) (see Fig. 4.6).

These systems have many diverse potential applications for telecommunications, broadcasting, crop monitoring, fire detection, tele-health services, education, policing or even defense-related services. The applications for a small island country, a state or province or other jurisdictions are just now beginning to be scoped out. Some have even suggested that HAPS-related markets could exceed $1 billion within 5 years or so. This new activity at the low end of the protozone is yet another option in the new commercial space world [5].

There are certainly a number of issues still to be addressed and resolved with regard to the use of the protozone. Addressing these will become more and more urgent as the proposed uses of this stratospheric area continue to increase. Table 4.2 presents not only the diverse applications that aspire to "fly" in the protozone but also the possible size of the markets. The chart below is of particular interest, not only because of the diversity of use but the great differences in velocities that various protozone vehicles would routinely represent, from virtually no movement to Mach 6 or higher. In terms of protozone flight safety this is a huge concern, given that there is no one responsible for traffic control and management of this "Wild West in the skies" protozone [6].

Table 4.2 Future commercial markets for systems that utilize the protozone (Image supplied by the author. All rights reserved.)

Projected 2035 Market Size for New Services Type of Service	Projected Revenues
Supersonic/Hypersonic flights in the Extreme Stratosphere (i.e. Protozone flights)	$10 billion/year or more
Commercial launches to low Earth orbit	Up to $10 billion/year
Space Tourism/Space Adventures via spaceplanes	$2 billion/year
High Altitude Platform Systems for telecom, remote sensing and surveillance and UAVs	$2 billion/year
Private Space Habitats for experiments and space tourism	$1 to 2 billion/year
Protospace robotic transport	Up to $1 billion/year
Dark Sky Station for experiments and ion engine launch of small payloads to low Earth orbit	Up to $1 billion/year

The Not Too Distant Future

In the next 5–10 years we may indeed have hypersonic jets that can fly across oceans in 2 h. We may have rocket motors that can be fabricated using 3D printers that cost 10–100 times less than those in use to today. The major supplier of launch services may be new companies that are merely start-up ventures today. The one constant in the New Space world is rapid change and innovation. Disruptive technologies is one of the constant themes that you will find in this book. Most people today equate space with rocket boosters. The roar of a rocket blast and vibrating force of a Saturn V launch as part of the Apollo Moon mission is something that is thrilling and impossible to forget if you were ever there.

However, the future is about change, and innovation still has far to go. In 20–50 years, the idea that the best way out of Earth's gravity well is to strap people on top of a controlled bomb explosion may grow to be seen as quaint or even primitive. It may 1 day be possible to use the Earth's electromagnetic field and its rotational or centripetal force to lift people and cargo at least to Clarke (geosynchronous) orbit. The use of tethers to lift mass into space, the construction of a space elevator or funicular to Clarke orbit, or even a lunar-Earth connector might 1 day make space travel three things it is not today: (1) much, much safer; (2) significantly lower in cost; and (3) a global utility that is available to all people and organizations that would aspire or have a purpose to go into space.

What many people do not realize is that the time-space continuum is something that works quite differently once one goes into space. When you get all the way out to the Clarke orbit, for instance, the pull of gravity is 1/50th of

4 Commercial Space Transport, On-Orbit Servicing and Manufacturing

that which applies at sea level, and from there it is possible to go almost anywhere in the Solar System without using huge amounts of energy.

You might use similar amounts of energy regardless of whether you are going to the Moon, Mars or somewhere else. The time to travel there, of course, can be greatly different, but the power or energy levels that are needed to be spent to get there are not greatly different. Also one can use the gravity wells represented by the planets and the Sun to speed up or slow down spacecraft, and this process, called "gravity assist," can allow enormous speeds to be attained by spacecraft. In short, once humans find a way to efficiently get as far away from Earth as geo orbit (at 35,780 km, or 22,270 miles) the rest of the Solar System becomes accessible. If you can get to a Lagrangian point, the entire Solar System can become our oyster. This is a long-winded way of saying that if we could ever build a space elevator to the Clarke orbit, or create a link to the Moon, we could go almost anywhere we want in the Solar System.

This is also another way of saying, the future of the New Space economy does not really depend on building a better, cheaper and safer rocket launcher system, but rather to invent the technology that would let us move easily and reliably—at least as far out as the Clarke orbit. After that a wide range of known technologies, from electronic propulsion to nuclear propulsion to solar sails and to gravity assist would allow us to at last navigate the cosmos. Just as the covered wagon gave way to the railroad that gave way to trucks and automobiles, which gave way to aircraft, which gave way to rockets, there is definitely newer, better, safer and more cost effective technology to replace today's explosive chemical launchers.

The Legal, Regulatory, Economic and Technical Challenges

The development of New Space transportation systems comes with a number of challenges. The first is the high cost. This is a challenge for space agencies, which have competing demands for the money in their annual budgets, but private aerospace companies and especially start-up ventures such as Rocket Lab, Firefly, XCOR, and even SpaceX, Virgin Galactic, Orbital ATK, External Engines and Bristol Space Planes often have to depend on space agency funding or very far-sighted angel investors such as Paul Allen to develop this difficult and demanding new technology.

Many issues with regard to space transportation systems will likely persist for some time to come. The top six issues that need to be addressed as soon as possible, however, can be summed up as:

Traffic Management and Control

This includes space and protozone traffic management and control—nationally and internationally—including protection of all radio frequencies associated with these operations. This may involve not only new international agreements but also new technical capabilities related to radar and tracking systems, precise navigation, positioning and timing satellite systems, etc.

Space Safety Standards and Controls

These would be for commercial space vehicles, for associated crews and passengers as well as strengthened controls for launch sites/spaceports and appropriate levels of safety oversight and certification or licensing.

Space Debris Remediation and Environmental Oversight

This would involve both new and improved methods to cope with space debris and its removal, as well as international agreement as to who should provide oversight with regard to air pollution monitoring and control for the stratosphere.

Internationally Agreed Processes for In-Orbit Servicing

This would involve agreed processes to handle on-orbit servicing, retrofitting and re-purposing of in-orbit spacecraft and other space objects.

Liability Protection and Settlements

This would involve seeking to mesh and coordination national legislation with international guidelines, standards and international agreements.

Resolving Conflicts and Overlaps Between Commercial, Governmental and Defense-Related Space Systems

This would entail a new global effort to harmonize commercial, governmental and defense-related space transportation system operations with regard to both space and/or protozone operations (including high-altitude platform systems (HAPS), dark sky stations, robotic freighters, hypersonic and suborbital spaceplane flights, space stations and habitats, governmental and defense-related flights and operations above current altitudes reserved for commercial aviation. etc.).

The number one starting problem, however, is that there is really no single international agency or body that has been afforded clear responsibility with regard to these various issues. At the international level the U. N. Committee on the Peaceful Uses of Outer Space (COPUOS), the International Civil Aviation Organization (ICAO), the World Meteorological Organization (WMO), the U. N. Environmental Program (UNEP), the International Telecommunication Union, and the U. N. Office of Disarmament Affairs are some of the agencies that have a legitimate interest in these subjects. The end result of having too many entities involved in oversight and regulation is that there is effectively no one with any real regulatory power at the stage.

The same, of course, can be said about national regulation and control of space-related matters. To take the United States as but one example, there are six Departments with some degree of involvement and assigned legislative authority. These include the Department of Transportation (FAA-AST), NASA, the Department of Defense, the Department of Commerce, the Department of State, plus the Environmental Protection Agency (EPA) and the Federal Communications Commission—not to mention the U. S. Congress, the President and the Supreme Court.

The recent Commercial Space Act of 2015 (HR 2262) made some attempt to clarify duties and responsibilities and assign specific agency roles, yet ambiguities still abound. The same is true for essentially all other spacefaring nations. The only thing that will likely change this condition seems to be actual military conflict among spacefaring, or "protozone-faring," countries. Such disputes will likely involve one or more of the above six areas of concern. It is in these arenas that significant conflicts or legal or financial claims will likely arise. Apparently only billion-dollar-level claims or an actual conflict in space could serve to force new agreements or at least negotiated settlements.

In the real world, national legislatures and heads of states tend to respond to crises and system breakdowns. Proactive action to untangle problems before conflicts occur is always what political scientists and advocates of international law espouse. But the chances of solving major commercial space-related problems beforehand remain minimal. Space businesses that are on a sound financial basis will likely somehow survive this lack of elegant regulation as the new gold rush continues to play out over time.

References

1. Mike Wall, "Private Orbital Sciences Rocket Launch Explodes During Launch" Space.com, Oct. 28, 2014. http://www.space.com/27576-private-orbital-sciences-rocket-explosion.html.
2. Jeff Foust, "Commercial space advocates remain confident despite accidents," Space News, Nov. 3, 2014. http://spacenews.com/42412commercial-space-advocates-remain-confident-despite-accidents/#sthash.t9xJPU1a.dpuf.
3. Chris Berin, "SpaceX Falcon 9 failure investigation focuses on COPV strut," NASA Spaceflight.com, July 20, 2015. http://www.nasaspaceflight.com/2015/07/spacex-falcon-9-failure-investigation-focuses-update/.
4. Chris Bergin, "SpaceX plans to debut Red Dragon with 2018 Mars mission," NASA Spaceflight.com, April 27, 2016. https://www.nasaspaceflight.com/2016/04/spacex-debut-red-dragon-2018-mars-mission/.
5. Peter Selding, "Thales Alenia Space Wins Initial Funding for High Altitude Platform, Plans 2018 Demo (April 26, 2016). http://spacenews.com/thales-alenia-space-high-altitude-platform-wins-initial-funding-plans-2018-demonstration/.
6. Joseph N. Pelton, "Beyond the Protozone: A New Global Regulatory Regime for Air and Space, The Manfred Lachs Conference, McGill University, Air and Space Law Institute, May, 2015.

5

Solar Power Satellites and Space Mining

Introduction

Space mining and solar power satellites are not science fiction. These enterprises are moving from dreams to experimental tests and technology development to the formation of actual businesses that are now seeking to implement these new resource-capturing capabilities in space.

In short there are real companies with real employees, raising real capital to support actual ventures that want to bring new assets to a resource-starved world. These vital New Space commercial activities will be keys to replacing fossil fuel energy and assisting with natural resource shortages that will become increasingly common by the end of the twenty-first century. Without the ability to access space resources our planet and the global economy as we know it could wither away under the weight of too many people and too little resources. Even with a transition to longer term sustainable practices and a world with zero population growth there are likely to be gaps in available resources.

In short, the New Space economy is the pathway to the future as we enter a new era for humankind. Within a half century, this New Space era could be as important as the industrial revolution was to the global economy in past centuries. But this transition must be done with intelligence. If it is not done correctly, the longer-term sustainability of the planet and of human civilization will remain at risk.

These New Space enterprises must do more than simply replenish diminishing natural resources and fossil fuels. They must be broadly conceived as part of an overall strategy to transform human society. There must be an integrated concept as to how we will cope with climate change, develop a global economy powered by clean energy and space systems that effectively deal with more than power and resources from the cosmos. In particular, there must be complementary strategies to address the problems of space debris, and even evolve new strategies to cope with climate change, cosmic hazards such as coronal mass ejections and asteroid strikes on Earth.

The New Space economy thus includes a wide range of new concepts for the global economy that include sustainability, recycling, limiting excesses, and evolving a kinder and more equitable approach to global society. The New Space economy can help steer us toward a new ethos and even a re-envisioning of the life styles of millennials and their children and their children's children.

The Business Logic and Technology for Solar Power Satellite Systems

The idea of launching and operating a solar power satellite is far from new. Peter Glaser of Arthur D. Little, the father of solar power satellites, was on the "Cruise beyond Apollo" in 1972. This cruise on the S. S. *Statendam* was a destination voyage to watch the night launch of Apollo 17, the last launch to the Moon. Following on to this amazing night launch off the coast near the Kennedy Space Center, Carl Sagan gave a tour of the Arecibo radio telescope in Puerto Rico. Some of the great scientists of the day were on board, along with science fiction greats Robert Heinlein and Isaac Asimov. This author was also aboard, at the very start of my career, and this was totally fascinating opportunity to hear from some of the giants in the field of space.

Peter Glaser explained in detail how it would be possible to deploy a solar power satellite that could beam power back to Earth at needed locations 24 h a day. The big debate on the cruise in 1972 related to solar power satellites was whether the transmission should be via laser or radio frequency transmission. English scientists argued for lasers, and Peter Glaser maintained that we should reserve radio frequencies for this task [1].

In the half century since Peter Glaser developed the first details about how solar power satellites could distribute clean power to the world, a great deal of new technology has been introduced. We now have better and lower cost space transport, better photovoltaic cells and perhaps soon have quantum dot technology to further boost performance. In short we have much better ideas about how solar power satellites and their ground receptors could be designed, developed and even operated profitably if the systems are designed with a sufficiently long lifetime.

The new commercial space transportation companies have developed a range of new systems that can lift satellites reliably and at significantly lower cost. New Space billionaires Elon Musk, Jeff Bezos and Sir Richard Branson have given us SpaceX and the Falcon 9 Heavy, Blue Origin and the New Shepherd launch system and Virgin Galactic the Launcher One, and these are just a few of the innovations that make space transport and orbital missions much more viable today. Arianespace is fighting back with Ariane 6. India is even developing a reusable space shuttle that may prove to be extremely cost effective and accomplish the regular cost-effective trips to space that NASA's space shuttle was never able to truly deliver.

Only if it is cost effective to launch solar power satellites into the skies will this New Space industry thrive and deploy solar power satellites on any significant scale. Innovative new launch systems and New Space industries may make this possible.

The second critical technology is, of course, more efficient photovoltaic cells, ones that are much more effective at converting solar energy to power. This new efficiency is being accomplished in part by capturing energy in the ultraviolet range as well as converting energy from across the visible light frequencies. Further, the new quantum dot technology, which is now thought to be less than a decade from commercial availability, promises even higher efficiencies—perhaps exceeding 50 % efficiency in converting solar radiation to usable power. There is also important research on protective coatings on the solar cells so that P/V systems can stand intense radiation for longer periods of time.

The most significant developments may be in innovative new architectures that allow the design of highly efficient and lightweight solar power systems in space. Key to these new architectures are the design of very low mass solar concentrators that would allow the solar power PV cells to "see" or effectively receive the equivalent power of not one but perhaps hundreds of suns. Such space-based concentrators can be of very low mass and of much higher efficiency than ground-based systems because of the three-dimensional flexibility of space and a weightless environment. In particular, there is the ability to create high levels of concentration without regard to radiation levels that would create health concerns on Earth's surface.

Some of the new solar power satellite designs envision three component parts. First, there are the lightweight solar concentrators that focus the equivalent of many suns toward the solar energy converters. Second there are the P/V cells or quantum dot units with glass protective coatings designed to preserve their lifetime—perhaps for decades. Third there are the RF or laser transmitter systems that would beam the energy back to Earth. In addition

Fig. 5.1 NASA design concepts for solar concentrator systems to create the equivalent of 100s of suns (Image courtesy of NASA.)

to these three components of space-based systems there is also the need for a ground-based reception system known as a rectenna [2] (see Fig. 5.1).

The design and location of the ground receiving system is highly critical. The receiving units must be distributed over a large area so that the beamed power is not, in effect, a super-intense "death ray" microwave beam. The ground-receiving rectennas would need to be distributed over a wide area measured perhaps in square kilometers to lessen the intensity of the downlink beam so that passengers on an airplane passing through such a beam would not be "nuked" like they were in a microwave oven. This receiving rectenna would also need to be carefully designed to prevent energy being reflected back into space. It is possible that return reflections could create unacceptable interference to sensitive communications satellites and precise positioning, navigation and timing satellites. Very strict safety standards have been suggested for this reason. Some of these safety standards for the maximum intensity of power at the rectenna's center might be around 25 mW/cm^2 (less than one quarter the solar irradiation constant). Further, the intensity on the outside edge of the rectenna zone has been proposed to have an exceedingly low level of less than 1 mW/cm^2. This translates to 250 W/m in the center and 10 W/m at the edge. This would translate into an area of over 10 million m^2 or 10 km^2 (3.8 sq. miles) to receive something like 1.5 GW of energy.

Only if the receiving rectenna was located off shore in the ocean, over a desert area, or in previously mined areas would this large amount of area be easily available. Thus it would seem that higher energy intensities would likely be required for the economics of solar power satellites **to** work. Figure 5.2 below provides a basic schematic of a large rectenna for receiving a continuous flow of energy from space in a desert area.

It is, of course, possible to conceive of systems that would distribute smaller amounts of power to smaller rectennas at different locations, but even if we were to think of sites that received just 250 MW of energy this would still require rectennas that were about 2.5 km^2 (1 sq. mile) in area. Currently Japan has championed the use of 5.8 GHz as a suitable radio frequency for downlinking microwave energy. This frequency range is close to the frequency used for non-licensed industrial, scientific and medical (ISM) applications and for C-band satellite communications.

The Challenges of Solar Power Systems as a Business

The challenge of solar power satellites as a business is complicated. There are the technical and operational challenges that are often related to costs. Key questions abound. What is the most cost effective space transportation

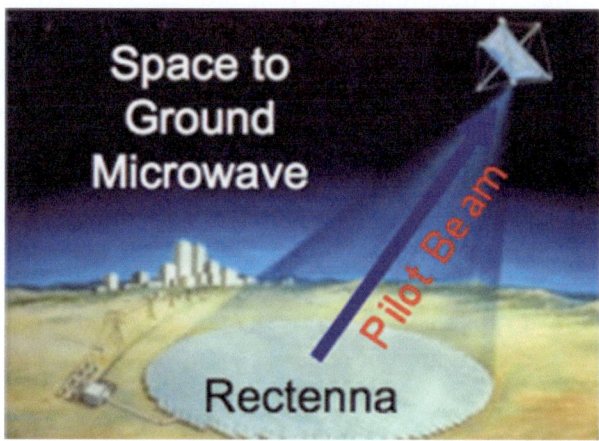

Fig. 5.2 Giant ground rectenna for receiving power from a solar power satellite (Image courtesy of Wikimedia commons. https://en.wikipedia.org/wiki/Space-based_solar_power.)

system to deploy a solar power system? What is the best technical design? Does this mean more effective solar concentrators? Higher performance solar energy conversion P/V/ cells? Waiting for quantum dot technology? Systems that have longer life? Better RF or laser transmission systems? More effective design of ground-based rectenna systems or systems that are distributed to more locations? What about the costs of the large area where ground-based rectennas are to be deployed?

Then there is the issue of competitive clean energies that do *not* involve space travel and which seem to be developing on many fronts. The new and emerging options seem almost limitless. There are wind farms, solar cell systems on the ground on homes and buildings, geothermal energy, ocean thermal energy conversion, ocean current turbines, compact nuclear fusion, Chemically Assisted Nuclear Reaction (CANR), Low Energy Nuclear Reaction (LENR), energy from silanes and hydrosilicons and so on. Some of these sources, such as geothermal and ocean current turbines—just like solar power satellites—can produce energy on a 24/7 basis.

There has yet to be a definitive case made that solar power satellites can, on a long terms basis, provide reliable and cost competitive services that are economically superior to these other alternative energy sources. If the price of hydrocarbon fuels rises substantially, the performance of solar cells improve significantly, and/or the lifetime of these SPS facilities increase substantially the cost equations change. In the next two decades the cost equations will change and perhaps substantially so.

But then there are those that indicate that we must think outside of the box. There are indeed design studies that indicate that if we proceed with space mining it might be possible to fabricate solar power satellites from materials obtained from the Moon or asteroids. Figure 5.3 below represents such a solar power system derived from space mining, processing and fabrication operations in space [3].

This analysis, which was admittedly undertaken as part of the rationale in support of space mining ventures, argues that if space mining and space-based fabrication using 3D printing technology are all taken amazing results can be achieved—perhaps by or even before 2050. In these scenarios the future of space transportation, space-based energy generation, and space-based manufacturing and fabrications are not really separate ventures, but integrated and symbiotic activities that become a win-win-win type enterprise.

The one failing that you cannot attribute to the space visionaries of our times would be that of thinking too small. These space entrepreneurs, who are contemplating off-world industries, envision a whole entangled web of various space enterprises. They envision communications and transportation

Fig. 5.3 Artist's conception of a solar power system fabricated from materials mined from an asteroid (Image courtesy of Wikimedia commons. https://en.wikipedia.org/wiki/Space-based_solar_power#/media/File:Solar_power_satellite_from_an_asteroid.jpg.)

systems, energy systems, environmental systems, habitats, mining, processing and manufacturing systems that can work in outer space and ultimately sustain off-world civilizations.

Distributing Clean Solar-Generated Power to a Needy World

The focus of most studies of solar power satellites has been on how to beam down energy on RF frequencies, or perhaps less likely via light frequencies, on a 24-h-a-day schedule. The other aspect that has not received the same attention is that the power could, in theory, be beamed down to essentially any location that might need it on Planet Earth. Currently there are countries with developing economies that are heavily reliant on coal or even wood-burning systems to supply electrical energy. A space-based system could be designed to provide distributed ground systems to meet the needs of countries with inefficient power systems that generate greenhouse gases.

Many concepts focus on geostationary orbit to supply energy on a 24-h-a-day basis to a single country. If one were to position a solar power satellite at a location such as the L-1 Lagrangian point, however, it could provide energy to wide range of countries, although this would create the need for energy storage systems. This could actually lead to using a solar power satellite as a form of international aid to countries that would participate by creating their rectenna system to receive clean power from space.

Solar Power Satellites and Planetary Shield?

The Sun is central to life on Earth. Without the Sun, life on our planet would end. It is ultimately the source of all power or energy that we use. The ironic fact is that solar storms also threaten life on Earth. Without the geomagnetosphere and the atmosphere, current levels of solar radiation would cause major genetic mutations, and perhaps kill off most life forms as we know them. Further coronal mass ejections without the protective shielding of the Van Allen Belts as formed by the magnetic poles could perhaps destroy all of the power as well as the electronic, communications and information networks that we have built to sustain modern life.

As we create more and more elaborate and sophisticated energy grids, IT and communications networks, as well as pipeline distribution systems we actually become more and more vulnerable to massive solar storms. Actually, the probability of modern industrial society being wiped out by a massive coronal mass ejection is far greater than the likelihood of it being wiped out by a wayward space rock.

For years people said that things like trains, automobiles, airplanes, elevators, electrical power grids, nuclear power plants, rockets and artificial satellites would be impossible to build, but build them we did. Today people say that things like space mining, solar power satellites, and space colonies will be impossible to accomplish. But someday we will do these things.

Just another "impossible" thing on the list is the idea that we might not only build solar power satellites, but that we might go further and create a system in outer space that is a solar shield that could temper the effects of solar radiation and block the most destructive coronal mass ejections that would threaten our satellites, electrical power systems, and communications and IT systems. This may sound like "pie-in-the-sky" or Venetian blinds in the sky, but there is good reason to start thinking seriously about ways to protect ourselves from the Sun. The fact is that Earth's magnetic field is weakening, perhaps on its way to a flip of the North and South poles. A weak magnetosphere has a greatly reduced ability to protect us from violent solar storms. Perhaps this field will soon be only 15 % as effective as it once was. By 2040 we may need to have some form of "solar shield" in place. We will discuss in more detail in the chapter in this book about cosmic hazards.

How Close Are We to Actually Doing Space Mining?

And where do we stand in terms of truly developing a space mining industry? Sure there are companies that claim to be space mining companies, but are these more than smoke and mirrors on an Internet website? Can such ventures truly be viable? And if so, when?

The truth is, there is still a good ways to go. Currently there are four companies that are pursuing space mining and they are all U. S.-based. These startup companies are Planetary Resources, Inc., Deep Space Industries, Shackleton Energy and Moon Express. But these are not your typical startups operating out of a garage. These companies have some serious names behind them. Billionaire Naveen Jain is the co-founder and chairman of Moon Express and the former president of Infospace. Planetary Resources started with billionaire backers that included Ross Perot, Jr., Google Chief Executive Officer Larry Page and Chairman Eric Schmidt, and former Goldman Sachs Group Inc. Co-Chairman John Whitehead. Since then it has added Google board member Kavitark Ram Shriram and International Software Corp. founder Charles Simonyi. Top advisor to the project is noted Hollywood director James Cameron. Shackleton Energy Company has some backing from Texas and Norwegian energy companies and has formed a partnership with Zaptec, a Norwegian mining technology company. Deep Space Industries has a more complex business plan that involves not only the idea of space mining but also refueling of spacecraft and other space-based services.

There is clearly skepticism as to how soon space mining could really happen. Skeptics challenge the very idea. What possible resources could be cost effectively reclaimed from space that would have sufficient value to pay for the huge investment costs? What space resources make sense, given the competitive advantage of mining carried out in land mines and even the oceans?

In response to these critics, however, spokesmen such as Rick N. Tumlinson of Deep Space Industries has said their plan is to use their 6-unit cubesat Firefly prospector system to identify asteroids with water content that could be used "to produce liquid hydrogen and liquid oxygen rocket fuel in space so as to avoid the huge cost of bringing launch propellant up from the Earth's surface." As noted on the Deep Space Industries web page the first step in mining is prospecting. Thus their prospecting mini spacecraft will undertake a close approach to a candidate near Earth asteroid (NEA). This prospecting spacecraft, as it goes into orbit around the asteroid, will use "spectral imaging and other research methods" to determine if this is a prime candidate for space mining and perhaps move to another orbit more accessible for space mining. Additional data will also be relayed back to Earth. This information will help scientists better determine the size, shape, spin and composition of the asteroid [4] (see Fig. 5.4).

Others, such as Jim Keravala, founder of Shackleton Energy, in briefings he gave in Washington, D. C., and Montreal, Canada, has emphasized that one of the prime objective of his company is "to obtain rare isotopes such as helium-3" that are not possible to find on Earth but are accessible on the surface of the Moon." It is thought by Keravala and other scientists at Shackleton Energy that helium-3 could be used as a key fuel for nuclear fusion. The even more ambitious part of the plan developed by Shackleton Energy is to use the helium-3 to power spacecraft that would place in Earth orbit solar power

Fig. 5.4 Concept design for Firefly prospector space probe by Deep Space Industries (DSI)—the size of a computer (Image courtesy of DSI.)

satellites that would also be built from materials mined and processed on the Moon. This seems today incredibly visionary and incredibly ambitious. Yet, if someone had predicted after the launch of Sputnik in 1955 that there would be a humans landing on the Moon in 1969, their forecast would have also been dismissed as sheer fantasy.

Lockheed Martin scientists who are working on what they call their compact fusion system have indicated that their plans, however, currently envision the use of deuterium and tritium available from "heavy water" that is obtainable here on Earth.

One leading space mining zealot, Peter Diamandis, has yet another slant on space mining. He says that Planetary Resources' Akyrd-3R probe will be seeking out asteroids that contain not only water and volatiles that could be used for fuels but also asteroids that are extremely high in platinum. Diamandis believes that some of these asteroids that could be steered back to orbit the Earth or the Moon, and some of the asteroids out there might have an ultimate market value as high as in the hundreds of billions of dollars, based on current values. Of course that much platinum might affect the global market prices just a tad.

Today's space mining companies represent a diversity of viewpoints on several key points. Some envision mining on the Moon, and others are more focused on the mining of asteroids. The potential targets for these space mining operations also vary widely. Some focus on volatiles, and especially water, that could be broken down to hydrogen and oxygen to create space-based "filling stations" for rocket launchers. Others talk about obtaining rare substance such as helium-3 isotopes, and yet others talk about finding asteroids that are nearly pure platinum. Some are thinking of mining of space objects where they are, such as on the Moon. Others are suggesting that they might change the orbit of near Earth asteroids so that they would ultimately orbit the Moon. Most agree on the need for careful prospecting to find reasonable asteroid targets before actually thinking of space mining operations.

To some people asteroids are celestial bodies. But to those focused on space mining, the millions of asteroids that circle the Sun are just space junk that have no intrinsic value. In fact they see the literally millions of near Earth asteroids out there as being potentially very dangerous assault weapons that could do a great deal of damage. So why not mine them rather than let them destroy us?

Countries Potentially Interested in Space Mining

The fact that the four mining companies are all located in United States might suggest to some that it is only U. S. entrepreneurs that are interested in the possibility of space mining and that other countries are completely disinterested.

This is simply not true. The book *Space Mining and Its Regulation* (2017) provides an extensive review of all the prior space prospecting activities, conferences and workshops on space mining and other related undertakings that have been carried out by other countries around the world [5].

There have been space mining-related activities, conferences, and workshops conducted in Australia, Canada, China, at the European Space Agency, France, Germany, India, Israel, Japan, Russia, and the United Kingdom, among others. There are mining associations in Australia and Canada that have explored with some specificity how they could produce automated mining equipment that could operate in space. In addition to the U. S. law that seemed to legitimize space mining, Luxembourg and the United Arab Emirates have taken steps to do the same. Many of the space agencies that have undertaken missions to the Moon or other celestial bodies have included experiments or activities aimed at assessing the viability and identification of potential sites where space mining operations might be feasible.

There is widespread recognition that many metals, rare earth materials and other natural resources are becoming scarce on Planet Earth. There are also current planning activities to design future space ventures involving habitats on the Moon or Mars. These and other future space missions will need to engage in space mining to sustain such facilities. The Sternberg Institute, one of the participants in the Russian Luna missions, has indicated that there is particular interest in prospecting on the lunar surface to determine if there are sufficient rare earth metals for space mining. Vladislav Shevchenko, of the Sternberg Institute, has said that mining the Moon could be the solution to the current shortage of rare earth metals, particularly since the main supply of such metals is tightly controlled by China [6].

Similar expression of interest in space mining has come from Canada and Australia, where experience and equipment associated with robotic mining has been acquired. In fact most spacefaring nations that have given serious thought to dwindling supplies of rare natural resources and metals have expressed some level of interest in space mining and have conducted studies as to how this might be accomplished in future years.

International Legal and Regulatory Framework for the Solar Power Satellite Systems and Conducting Space Mining

Solar power satellite systems that beam electrical power to Earth and space mining activities of the Moon and asteroids are two likely future space industries. The time frame is hard to pin down, but many space experts feel that some of these industries will be actually be operational within the next 10–30 years. Both will involve some level of agreement on the legal and regulatory framework that will be needed to carry out these activities, but the issue of solar power satellites is far easier and straightforward to address.

In the case of solar power satellites, the issues that arise in space are few in number and easy to identify. These involve the appropriate electromagnetic frequencies that would be used to manage such systems and to downlink the energy transmissions without interfering with adjacent communications satellites or ground stations for satellite communications or creating health hazards in the area where the rectennas are operational. There would need to be standards to ensure that there is no reflected energy back into space that is strong enough to create unacceptable interference. The primary need is to establish standards to ensure that the SPS downlinks would not be too powerful and thus create health issues for people that live nearby or are in airplanes that would fly through the beams. Thought has been given to such issues, and the likelihood is that appropriate health and power emissions standards can be agreed on well before operational solar power satellites are launched. There are zoning questions as to suitable locations for large scale rectennas and there also environmental and ecological questions that will be addressed as well. In the case of solar power satellites it appears that technological and business economics will be the driver as to when solar power satellites will be manufactured and launched into space, and no larger issues of international law will be involved. This is not the case for space mining.

Most also agree that greater clarity is needed about the interpretation of national and international law concerning the legal status of space mining and the "dos" and "don'ts" that will ultimately apply to such activities. Clearly the U. S. firms discussed earlier (Deep Space Industries, the Moon Express, Planetary Resources Inc., and the Shackleton Energy Company) are heartened by the passage and enactment into law of the Commercial Space Act of 2015. This new law would seem to be the clearest indication yet that not only can space mining become real, but that a legal and regulatory structure is likely to be created to establish rules of the road or regulations of the skyways to allow these types of space activities to occur.

This law is certainly helpful in starting the discussion. Its stated purpose is to: "promote the right of United States citizens to engage in commercial exploration for and commercial recovery of space resources free from harmful interference, in accordance with the international obligations of the United States and subject to authorization and continuing supervision by the Federal Government" [7].

It also defines the activity of space mining as being able to remove from "outer space" only "abiotic" resources. This definition thus presumably applies to inert materials that do not include any life form and is not limited to any particular place in outer space. In short, it would seem to cover asteroids, near Earth objects, the Moon, planets and their moons, or even the Sun—indeed everything in the cosmos in an off-Earth location. Appendix 1 in this book provides the latest recommendation from the Obama White House as to how such space mining operations would be licensed and approved, even though other countries are contending that national governments cannot by themselves authorize space mining operations.

Clearly the U. S. law has already served to open the door for other countries to enact similar laws, as already noted to be the case for Luxembourg and United Arab Emirates. There are a lot of questions as to what will come next. Will the U. S. law be formally challenged or will more and more countries see that the precedent is much like the Law of Seas and Antarctica Convention. Countries or companies under this convention can engage in ocean mining without asserting sovereignty over the oceans. The president of Moon Express, Robert Richards, has said that this is no different than fishing in the ocean. Fishermen are not asserting sovereignty over the oceans and neither are we.

Many questions will follow. There will be a need to sort the efforts of specific space mining initiatives from one another. Will there be some sort of international process in addition to a national licensing process? What if two countries or private enterprises decide they wish to engage in space mining on the same asteroid or to move the same asteroid to a new orbit for easier access? Who approves or disapproves? There perhaps are as many new questions as might have been answered by this new act.

The list of questions that comes into play includes the following: (1) What does "in accordance with the international obligations of the United States" actually cover? (2) Is there a direct or simply implicit conflict between the new U. S. law and the Outer Space Treaty? What about the so-called Moon Agreement? Does the "Moon Agreement" now signed by 16 countries as of June 2016, apply to the United States in an international court even if the United States and most countries of the world have not ratified it? (3) Does outer space, like the oceans and Antarctica, represent a "global commons," and if so, what does that mean when it comes to mining any of these areas? (4) What does the Liability Convention mean in terms of private mining operations under the "licensed authorization" of a national government? What are the limits on the liabilities that might be incurred? If for instance a private entity moves a near Earth object to a new orbit and it ultimately smashes into Earth, or alters tidal patterns, what limits are there on the absolute liabilities that presumably apply to the country that licensed such mining operations? Can national governments really protect themselves against such private enterprise malfeasance or miscalculation? The list of unanswered questions is actually quite long.

The good news is that there may be at least a decade or two or even three before such questions convert from hypotheticals to real issues with real-world repercussions. The bad news is that this could truly open up a new arena of conflict between or among spacefaring nations. Instead of the United States and Russia seeming to be at loggerheads over Eastern versus Western Ukraine, it could be over space rocks in the sky or even the Moon.

The issue of space-related liabilities, and space miners staking claims in the sky today seems farfetched. But in time these seemingly hypothetical issues could become quite real. Pro-active discussions that follow the model used for the Law of the Seas and Antarctica could possibly help avoid future conflicts.

Today we must consider on how to set technical, safety and legal liability standards related to commercial and governmental space transportation systems. In time we will also have to develop safety standards and regulations for solar power satellites as well as space mining ventures. Current international tools include the Outer Space Treaty, the Moon Treaty, and the Liability Convention. The question is to find out if the Commercial Space Act of 2015 will be considered consistent with and a useful extension of international law—or not. Other countries will undoubtedly enact their own national laws governing such issues as space transportation, space safety, solar power systems and space mining. What is not clear is whether these are in parallel with U. S. legal precedent or directly contrary. Clearly there is a long ways to go. What we do know is there "gold" in the skies. It is in the form of clean energy, rare metals, and accessible volatiles that can be converted to rocket fuels.

Conclusions

The idea of space mining today still seems like science fiction to some. One early science fiction book, entitled *Rip Foster Rides the Gray Planet,* was about space mining. It excited the imagination of many and stirred thoughts about a distant future where astronauts might mine the Moon or asteroids or derive power from space. Then it seemed like this might take place centuries into the future. Now it seems only decades away, given the tremendous strides in technology that have allowed a great surge forward.

In the past half century there has been enormous progress in space transport, space habitats, and artificially intelligent robots that are capable of achieving progressively more demanding tasks. Forecasts of "smart robots" with machine intelligence comparable to humans have been projected to be less than a generation away [8].

In short the technology to create reliable and increasingly low-cost space transportation systems, small, low cost robotic "prospecting" spacecraft armed with sophisticated sensors, robotic devices with the "smarts" to carry out remote mining, and dozens of other technical capabilities to allow space mining are either currently being developed or are on the way.

There is no critical technology that cannot be made available within a decade that would block the development of actual space mining operations. One key question is whether humans will need to be available on site and supported by space-based habitats to do space mining or whether the entire operation can be achieved remotely via smart robots. Clearly solar power satellites can be remotely operated without humans present in space.

Another key question is whether operations that seek to mine near Earth asteroids will be accomplished in their current orbits, with mined materials or volatiles shipped back to Earth or a convenient location near Earth. The alternative would be to attempt to divert the entire NEO either into orbit around the Moon or Earth. The answers to these questions will not only drive the direction of technological advances but also have enormous consequences for those seeking to define the appropriate international legal framework and applicable regulations.

The largest question of all is who will enforce the new space laws and regulations that apply to space transportation systems, space mining, and other new space activities that will become more and more vital to the global economy within the next two or three decades.

When the airplane was first invented it took a while to recognize that someone should regulate and control these flights to insure public safety. With the invention of nuclear power plants and jet aircraft and international trade in food and drugs, the need for international controls and safety standards were recognized more quickly. It does not require great vision to recognize that today we increasingly will need controls for orbital space debris, hypersonic aircraft flying to the top of the stratosphere that would be operated as international transport systems, as well as agreed international standards and controls for solar power satellites and space mining activities.

References

1. Conversations between Peter Glaser of Arthur D. Little and Joseph N. Pelton, 1972.
2. Donald Flournoy, *Solar Power Satellites* (2013) Springer Press, New York.

3. Power Satellite from an Asteroid https://upload.wikimedia.org/wikipedia/commons/1/16/Solar_power_satellite_from_an_asteroid.jpg. Last accessed on Dec. 15, 2015.
4. Deep Space Industries web site, http://deepspaceindustries.com/prospecting/. Last accessed on Dec. 17, 2015.
5. Ram S. Jakhu, Joseph N. Pelton and Yaw Nyampong, *Space Mining and its Regulation* (2017), Springer Press, New York.
6. Ram S. Jakhu, Joseph N. Pelton and Yaw Nyampong, Space Mining and its Regulation (2017), Springer Press, New York. Chapter 3.
7. Commercial Space Act of 2015, Section 4, Section § 51302. HR. 2262, November 2015.
8. Ray Kurzweil, *How to Create a Mind*, 2012, Viking Press, New York.

6

Space Security, Defense and Weapons

Introduction

There is no doubt about it. Global defense today is closely wedded to space systems. Vital to today's strategic defense are a laundry list of space- or space-related systems. There are military communications satellites, weather satellites, navigation and timing satellites, satellite systems for surveillance, sensors to detect nuclear detonations, and satellites and ground-based tracking systems for detecting missile launches and space debris—or what military wizards call space situational awareness. This is not mention a whole slew of missile systems ready for instant launch from silos, trucks and submarines.

Billions and billions of dollars are spent on space-based systems as well as ground systems to track or navigate or support space systems. There are over 250 dedicated military satellites worldwide, or about 20 % of all the operational satellites. And this does not include the 50 % of all civilian or commercial satellites that have some dual use, or supplemental use for military purposes. Indeed some commercial satellites, such as like XTAR, are designed to operate exclusively in the military X band (8/7 GHz) and thus are indeed always used for military or emergency relief communications.

These space-based war-fighting systems in the skies are much further away from World War II than World War II is away the from Crusades. The new technologies that are being envisioned for space-based strategic operations are very high tech indeed. Although we did not deploy the strategic defense initiative known popularly as "star wars," most of the underlying technical capability is now available. Clearly the United States, Russia and China, at least, have anti-satellite capabilities and so-called "kill satellites." The same can be said about plans for the Transitional Satellite (TSAT) optical ring that would have been the equivalent of an optical cable systems in the skies but was canceled by Congress as being too expensive. Expensive is a relevant word, because some of the more sophisticated military satellites cost in the billions of dollars. And the related ground systems don't come cheap either. The new S-band radar system being installed in Micronesia by the United States to track missiles and space debris, being built by Lockheed Martin, will cost $7 billion, and there are tentative plans to create a similar system in Western Australia. The bottom line is that military defense systems from drones, missiles and satellites are being more and more automated, deployed more and more in the skies, and cost more and more money [1].

Almost 60 years of exploration and use of outer space have brought unprecedented benefits to humankind, but it has also brought a huge dependency on space as well. So-called economically advanced countries and war-fighters, in particular, now heavily depend upon space. And the future seems to be even more of a space-based investment. Even a single day without satellites would have disastrous impacts for everyone on Earth, particularly those who increasingly rely on space assets for military defense.

In 2014, the global space economy grew by 9 % (compared with 2013) and reached $330 billion in annual economic activity. The numbers of 2015 are expected to reach around $350 billion. Space products and services are not only a significant number by themselves, but even more importantly they have become indispensable for such applications as agriculture, climate change monitoring, communications, international arms control and disarmament, medicine and health care, banking, natural disaster management, transportation, and weather forecasting [2].

There are many economic, social and cultural gains this amazing array of new space activities now bring to our planet. Satellite applications have a tremendous impact on communications, broadcasting, networking, remote sensing, weather prediction, navigation, and much, much more. Space—far from being a "wasted expenditure"—now provides a vital contribution to global economic growth. Each year new space infrastructure and services become even more important.

As space systems become more economically central to commercial growth and also provide more capabilities to countries around the world, they are, in parallel, also becoming more vital to defense and security. The bottom line is that space-related defense systems today rely on space for many things. Key dependencies include communications, surveillance, navigation and targeting, protective shielding, and, of course, defensive and offensive weapon capabilities. Today these weapons systems are missiles, but there are new capabilities under study that include high energy lasers and directed energy systems, or even more exotic weapons systems called "Rods from Gods" that would send metallic rods at hypersonic speeds to kill adversaries. And there have even been studies of nuclear explosions in space to create electromagnetic pulses (EMPs) that could disable communications, computers, and even electronics in cars and aircraft. The march of technology and warfare is a very much arm and arm affair.

These defense-related space expenditures are increasingly on the rise for most countries with active space programs. The U. S. Defense Advanced Projects Research Agency (DARPA) that gave us the Internet and many of the advanced space systems in use today have studies afoot to build and service satellites in space and other projects that are too classified to discuss.

And then there is the other side of the coin. What national security issues could flow from a major attack on the space systems of developed economies? Just how vulnerable are countries that now have an enormous reliance on space systems? Ironically, the more advanced a country is today, the greater its potential vulnerability. As we will mentioned earlier this vulnerability is not only from a manmade attack but to cosmic events such as a massive solar storm.

What if there were a cyber or physical attack or huge solar mass ejection that affected—not in a good way—all the parts of the civil space infrastructure now up in space? The loss of space navigation and timing satellites alone could shut down much of aviation safety systems, and time synchronization of the Internet would be lost. The bottom line is that we are dependent on space systems to defend us, and if our civilian space systems were somehow knocked by an electro-magnetic pulse (EMP) from a nuclear bomb or a massive solar storm we would likewise be in deep trouble.

The truth is that a loss of space systems—either civilian or defense—would be enormous hits to the economies of the United States, all of Europe and the U.K., Japan, Canada, Australia, and even China, India, Russia, the Republic of Korea, Brazil, Thailand, Taiwan, Malaysia, Indonesia, Argentina, Chile, Colombia, Peru, Equator—virtually all of the countries of the Middle East and many countries in Africa as well. Perhaps a hundred countries would experience something between a solid gut punch to a virtual knockout. And the rest of the countries of the world would still have some ill effects. Again, it is useful to see the video on You Tube entitled "If There Were a Day without Satellites." This short video gives you a feel for all the things that would go wrong should if just communications satellite and position, navigation and timing networks should be knocked out [3].

The loss of so-called civilian use satellite networks could have a major impact on countries with an advanced service economy. In short, space systems are now vital to the national security and to the day-to-day economic functioning of many nations. This is true not only in terms of defense-related space activities but the overall economy as well. Spacecraft applications are now essential to day-to-day operations of most of the world economies. Thus, even an attack on commercial or civil space systems for communications, navigation and timing, weather forecasting, basic governmental functioning, transportation, banking, news and entertainment, etc., would be considered a major act of aggression. The problem is that it may not crystal clear to the major spacefaring countries whether or not this was a true act of war. Clarity on this issue is just one of the strategic issues that needs to be absolutely clear, because "killing a satellite" is not something that can be undone.

The line between satellite systems associated with military operations and defense on one hand and a diverse range of space systems used for civilian governmental operations and commercial enterprise has now become so blurred that a clear distinction or separation of the two is no longer possible. Dual use of commercial systems to support military operations is now pervasive. The civilian and commercial satellites are now central to the global economy. Such functions as aircraft takeoff and landing, global news and entertainment, global banking, operation of shipping and railroad management, as well as many information and communications systems, are all now dependent on satellite networks. Just knock out the synchronization of the Internet, and major corporate intranets and the global economy plus the national security systems of many countries would be clobbered big time. A third of all ATM banking systems in India for instance are dependent on satellite connections.

New capabilities developed for peaceful purposes within commercial space systems could also be deployed as part of defense communications, surveillance, and targeting. In some instances even dual use of commercial satellite system might be considered part of a national space weapon system. Communications to support the operation of drone systems armed with weapons are sometimes routed via commercial satellite networks. Normally such tactical communications, however, would be routed via defense satellite networks. The truth is that the overlap situation between commercial and defense-related satellite networks will only get worse. Global security systems are now tied to defense satellite networks as well as civil and commercial satellite networks in literally hundreds of ways.

New Regulatory Issues in Low and Middle Earth Orbit

The "high frontier" of space systems and the development of new technology will likely become even more vital to national security defense systems in coming years. Twenty-first century concepts of warfare and national protective

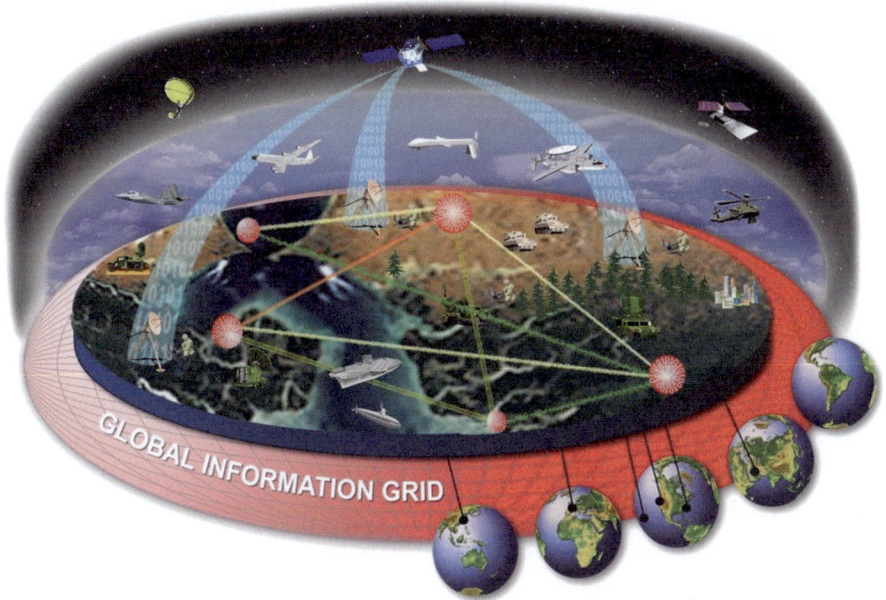

Fig. 6.1 The U.S. Global Information Grid (GIG) that heavily depends on many different types of communications satellites (Image courtesy of the U. S. Department of Defense.)

systems are more and more dependent on advanced broadband communications, satellite-based remote sensing, and positioning, navigation and timing networks. The image above illustrates some of the complexity of the U. S. Department of Defense's worldwide network to maintain instantaneous communications with all parts of Earth, including the oceans. The ability to tie this all together depends on communications satellites [4] (see Fig. 6.1).

As new space industries involve such areas as space mining, space manufacturing and processing, and space power systems, the ability to discriminate between civil/commercial space activities and military/defense-related space activities will very likely become even harder to sort out.

The future of space systems and national security threatens to become more complex for several reasons. One reason is that the very "dimensions of the space theater" we are talking about is growing wider both at low Earth orbit and above.

Suddenly there are more and more activities that are being carried out in the so-called protozone. As mentioned earlier, this protozone (also called protospace or sub-space) region represents the area above commercial airspace where aircraft and jets now safely fly under the control of international and national air traffic management and control systems. This zone is shaping up as a potential conflict zone. The protozone can be defined as the region above (21 km or about 13 miles) and below (160 km or 100 miles) the region where satellites can no longer maintain orbit From a regulatory perspective this region is now an area much like the Wild West of the old United States some 200 years ago. At that time there were really few, if any, marshals, sheriffs or law enforcers in town, and the judges were nowhere to be seen except when they came around for periodic visits.

This protozone is a difficult place from a regulatory point of view. It is a very broad region comprised of billions of cubic miles where there is a lack of agreed competent technical tracking and control systems at work and no single regulatory agency in charge [5].

This is a very vast area not effectively covered by sophisticated radar systems, and even precise pointing, navigation and timing satellites have some accuracy issues. This is definitely a region where there is a lack of agreed regulatory oversight. The new S-band "Space Fence" radar will, however, add a great deal of capability in this region on the part of the United States, but few other countries would likely wish to spend some $8 billion U. S. dollars to duplicate this type capability.

The regular use of the protozone is suddenly being proposed by all sorts of people and with highly different types of technologies and systems. There are plans for flights by spaceplanes, hypersonic point-to-point transportation vehicles, high altitude platforms, dark sky stations, robotic freighters, and perhaps even various types of military systems. This region is the "unsettled bottom." [6]

Then there are all sorts of thoughts about new types of human and robotic activity above Earth orbit. There are plans for space mining of the Moon and asteroids, building or creating space habitats, space systems, or even space colonies at off-world locations such as the L-1 Lagrange point, on the Moon or in lunar orbit, on asteroids, and even on Mars or in Mars orbit. This incredibly vast region is where there are no agreed international treaties and conventions, no established enforcement or policing authority. To date, the history of humankind shows rather dismal results in such situations. The lack of control and enforcement capabilities usually creates great opportunity for turmoil and conflict. Why should the protozone, orbital space, and true outer space well above Earth orbit prove otherwise?

Currently, when one sees the conflicting agendas between the United States and Europe grouped on one side and Russia on the other with regard to the Ukraine and Syria, the danger if clear. Areas such as the Ukraine, Syria, Yemen, Afghanistan, and Iraq are clearly defined locations here on Earth, where supposedly clear national and international agreements, peace treaties, and firmly established national government jurisdictions are in place. Yet vying spheres of political influence and armed conflict rages around the world.

If this is the situation on Earth, where the rule of law within a twenty-first century world is supposed to be in place, one must be fearful of the future for outer space. One can only imagine the possible conflicts that can evolve in the "Wild West" territories represented by the expanding zones of human influence in all three areas—within inner space, i.e., the protozone, within various Earth orbits, and above Earth orbit in true outer space above the world's gravity well.

Outer Space Security in a Twenty-First Century World

Thus, many concerns exist today with regard to the protozone. These include the exploitation of space natural resources and national security activities in outer space and even about new systems to be deployed in Earth orbit, especially since there are new proposals to deploy hundreds if not many thousands of satellites in gigantic constellations for new applications such as broadband communications. Today we have to consider not only the wide ranges of human activities in space that might give rise to conflict, but on top of other thing else, there are the perils of cosmic hazards from natural causes, such as from asteroids, comets, extreme solar events, and now a new and increasing threat from orbital space debris.

The spectrum of issues involving space law and security is quite broad. There are very few agreed international laws that are clearly "settled law." Since space activities are now quite dynamic, with new technologies and applications constantly cropping up, the regulation of space development is tremendously challenging. Space security is perhaps the most challenging of all.

This chapter thus does not attempt to address all of the possible space security issues that will likely arise in coming years. Instead it focuses on the "hot button" issues that seem most likely to erupt into disputes between nations and give rise to legal or regulatory problems in the foreseeable future. Those space security issues include: (1) Disputed use and regulatory authority over the protozone and the currently increasing need to create a "space traffic management and control" capability for this area; (2) Disputes that involve jamming, disabling or harming satellites, spacecraft, or any space system deployed in space by any spacefaring nation; (3) Disputes about who can engage in space mining activities and the ownership, selling and use of space-mined materials and substances either in outer space or when returned to Earth; and (4) Deployment and use of space systems that might be considered a weapon or space military systems by either states or even "space terrorists." We also address the special case of systems that might be used to defend against a natural cosmic threat or orbital space debris.

All of these issues might ultimately involve liability claims by one country against another, or by countries acting on behalf of one or more commercial enterprises. These liability issues and the meaning and effect of the Liability Convention have been discussed in some detail in other places and thus will not be repeated here. Only if future international agreements assign a role to private concerns rather than to nation-states does it seem that there will be any major shift in the international liability claims process related to space systems.

Urgent Security Concerns in the Protozone

The control and regulation of commercial airspace by national and regional aviation agencies and by the International Civil Aviation Organization under the Chicago Convention, amended since 1945, is well established. This covers civil air space that is generally understood to be the ground to an altitude of 21 km, or about 13 miles [7].

The Outer Space Treaty of 1967, formally known by the rather longish title of the "Treaty on Principles Governing the Activities of States in the Exploration and Use of Outer Space, including the Moon and Other Celestial Bodies," plus its associated treaties, conventions and international agreements on space-related matters all relate to what is rather vaguely called "outer space." Most of these formal treaties and conventions were accomplished by means of the U. N. Committee on the Peaceful Uses of Outer Space (COPUOS). Other agreements have been reached by means of the good offices of the U. N. Office of Disarmament Affairs (UNODA), or other collaborative means, and then agreed within the U.N. General Assembly. This U. N.-led process has allowed key international understandings to be reached with regard to the prohibition of space weapons placed into orbit or outer space, nation states' liability responsibilities with regard to space objects, and precautions that should be taken with regard to nuclear power supplies, etc. This is the good news.

The bad news is that there is currently a rather large gap in global space governance in what was earlier described as the protozone. This rather considerable gap exists in spite of increasingly ambitious plans to fly spaceplanes, hypersonic transport vehicles, high altitude platforms/aerostats and perhaps other craft within these altitudes, where there are no national or international regulatory bodies with a clearly defined responsibility for this region.

Currently there are many new applications for the protozone that include or could soon include: (1) suborbital flights on space planes—crewed and non-crewed; (2) hypersonic transportation—crewed; (3) stratospheric balloons and dirigibles—crewed; (4) dark sky stations and ion engine propelled craft; (5) robotic aircraft transporters; (6) high altitude platforms for communications, remote sensing, etc.; and (7) stratospheric surveillance platforms; and perhaps (8) various types of military surveillance, and crewed and robotic weapon systems and devices [8].

There would seem to be a most urgent need to address the uses of the protozone in the context of: (1) national or international traffic control and management (i.e., in relation to the technical, operational, and regulatory concerns over safety and potential collision avoidance); (2) allowable and prohibited uses; (3) national, private entity or international liabilities and responsibilities related to this now unregulated area; (4) additional safety, traffic routing and coordinative responsibilities and duties; (5) inspection and control beyond traffic control and management; and especially to define the status, control and prohibitions against the use or flight of weapon systems in this region.

The complex nature of current or proposed uses for the protozone seem only to become more and more challenging as the regulatory gap in this now vital area continues to grow. The current issues that seem to require urgent attention include liability as well as possible agreement on prohibitions against the use of weapon systems and military-related crewed and robotic devices in this region. Perhaps the most urgent matter of all is that of traffic control and management issues, both with regard to defense, governmental and civic applications for this now largely unregulated zone. Today there are active development programs in the United States, Japan and Europe to develop hypersonic spaceplanes. These development programs involve both governmental programs and commercial New Space companies.

Finally there is a question as to whether various types of uses in this region might somehow be licensed or at least formally acknowledged if for no other reason than to avoid collisions and to maintain safety in this region that seems destined to be used with ever greater frequency [9].

Licensing, Launching and Operation of Space Systems and Coordination of Radio Frequencies in Earth Orbit

Unlike the case of the protozone, there is currently some 60 years of experience with regard to the allocation of the use of radio frequencies, coordination of RF usage to minimize interference between space systems, registration of the launch of space objects, and other forms of regulation and coordination of objects launched into Earth orbit. There are now recommended procedures to minimize space debris and guidelines for de-orbit of space objects within 25 years of end of life. The International Telecommunication Union plays a prime role in most of the coordination procedures. In general most of the allocation procedures and coordination of space systems have worked reasonably well.

Nevertheless a number of issues have come to the fore in recent years. These include: (1) A significant increase in jamming of satellite communications systems, both civil and military systems; (2) new concerns about cyber-attacks on spacecraft as well; (3) The continuing deployment of a large number of small satellites (some of which are not registered under the Registration Convention) and newly emerging plans to deploy large-scale satellite constellations involving perhaps thousands of satellites. Such gigantic constellations and unregistered small satellites that could potentially lead to problems of runaway or cascading orbital debris build-up that would jeopardize defense systems for monitoring nuclear explosions, for targeting of missiles, for defense communications satellites and surveillance, etc.; (4) A surge in demand for broadband terrestrial mobile cellular service that is undercutting allocations for satellite communications, including military communications satellite systems; and (5) new uses for Earth orbiting satellites such as solar power satellites and high altitude platform systems that could interfere with or aid intersystem coordination for satellite networks, including military communications satellite systems.

Currently the International Telecommunication Union (ITU), a specialized agency of the United Nations headquartered in Geneva, Switzerland, performs the prime functions related to radio frequency allocations, intersystem coordination, coping with problems of interference and jamming, and creating new standards [10] (Fig. 6.2).

This institution operates on the basis of consensus agreement and has no true enforcement powers such as that provided to the World Trade Organization. The ITU operates through the power of official national administrations. If it is

Fig. 6.2 The ITU headquarters building in Geneva, Switzerland (Image courtesy of the ITU.)

the national administration that is engaged in jamming, not abiding by recommended standards or engaging in dangerous practices of not removing space objects that are space debris from orbit, neither the ITU nor the United Nations have enforcement powers or "space police" to address these problems.

Currently 23,000 space objects in low, medium and geostationary orbit are being tracked that are larger than a baseball. Of these less than about 12 % are operational spacecraft. Two collisions within the past decade have increased the number of tracked debris elements by about 5000. If another major collision occurs the number of major debris elements will increase by perhaps 2000–3000 pieces.

Problems that need to be addressed in the next few years include interference, jamming, orbital spacecraft use and congestion, space object debris build up, more competing civil and defense-related uses of the protozone, new competition for the active use of Earth orbits. Competition between terrestrial, protozone and Earth orbits will likely create the need for some new enforcement powers such as fines and perhaps the ability to remove offending space objects and/or debris from orbit. The problem is that there is no international agency with any kind of enforcement rights, and national actions to address such problems could lead to major disagreements between affected nations and might possibly be seen as an act of war. As systems in Earth orbit or the protozone—especially those associated with national defense—are seen as more vital, as frequency interference or orbital debris increases, or other types of conflicting use of systems in Earth orbit increases, the possibility of disputes among spacefaring nations can and likely will increase.

Further, the vital nature of space-based "civil infrastructure" just keeps growing. The U. S.-owned and operated GPS system alone is now key to aircraft safety, monitoring of nuclear explosions, synchronization of the Internet and the routing of cars, trucks, ships, as well as the targeting of missile systems. An attack on this system would clearly be an act of war.

Today a growing array of space objects and satellite networks have become as central to national defense as aircraft carriers, air force fighters or bombers, missile defense or any other aspect of a nation's military/defense infrastructure itself. In light of the "dual use" of civilian or commercial communications and remote sensing satellites for military uses, the distinction between civilian and military applications becomes more and more difficult to distinguish. Deliberate jamming or a physical assault or even a cyber-attack on vital satellites is now becoming a very serious matter. Some countries are now overtly declaring in their national space policies that such attacks are explicitly defined to be "an act of war" akin to invading another country's territory or bombing one of their cities.

The cumulative effect of all of these issues and potentially increasing and conflicting use of Earth orbit makes it clear that space security issues are mounting. Ultimately, it is apparent that the possibility of armed conflict in space is growing, and the general public remains largely unaware of these developments, even though they might have devastating implications should so-called "space wars" actually materialize. Therefore, in order to avoid potentially devastating conflicts and to regulate the military activities of states (and non-state actors) in outer space, there is a manifest need to clarify the applicable rules of international law and emerging codes of conduct, particularly rules governing the prohibition of the use of force, as well as applicable rules of international humanitarian law.

In the area between Earth's surface and up to at least up to geostationary orbit, there is currently a modicum of cooperation, standards related to the use of radio frequencies, and intersystem coordination managed within the processes of the International Telecommunication Union. For all of the reasons note above, however, this could become increasingly difficult. But what about out beyond the reaches of Earth orbit and new attempts to acquire resources that are in what is called "deep space"?

The Issue of Space Mining as It Relates to Space Security

On November 25, 2015, the United States adopted a new act that was intended to entitle American companies to private property rights with regard to natural resources they would mine in outer space. This act, H.R. 2262, the U. S. Commercial Space Launch Competitiveness Act, is known more simply as the Space Act of 2015 [11].

This has already aroused several concerns (or possible conflicts) internationally, primarily because the mining of asteroids and other celestial bodies in outer space could generate billions, if not trillions, of dollars in revenue and could also have serious implications for the space environment and the long-term sustainability of space in proximity to our planet. This act also called for a detailed study of space traffic management and further development of low-cost space transportation systems that could propel forward all sorts of New Space activities, including space mining.

The precedent that this new Space Act of 2015 (H.R. 2262) has set will put in motion actions by other countries. One expected course of action is for a number of other spacefaring nations to adopt parallel legislation to put their country on the same legal footing in terms of their ability to engage in space mining and to recover and use space assets. The other possible course of action would be to formally object in international courts to the legality of the new U. S. law [12].

Given the increase in the number of states and non-state actors that are now active in space, and the military's increased reliance in countries that have space capabilities, there are growing concerns about the risk of a conflict in outer space, or at least the possibility of incidents involving the disputed use of outer space systems leading to major confrontation. As space infrastructure grows more vital to global economic, business and strategic systems, the potential of space conflict appears to increase. Space mining alone may give rise to many points of conflict that go beyond competing for resources. If future space operations, for instance, should involve the redeployment of asteroids to lunar or Earth orbit, this could lead to very serious concerns about disastrous collisions and space safety. Likewise if one establishes a "temporary site" for space mining under a "national license" it certainly becomes the moral equivalent of planting a flag. It becomes the twenty-first century of creation of the East India Corporation by Britain (Fig. 6.3).

Commercial space activities seem likely to continue to grow apace. Space commerce may indeed ultimately grow to become a trillion dollar enterprise—i.e., truly the new gold rush. In such a future that may not be too dis-

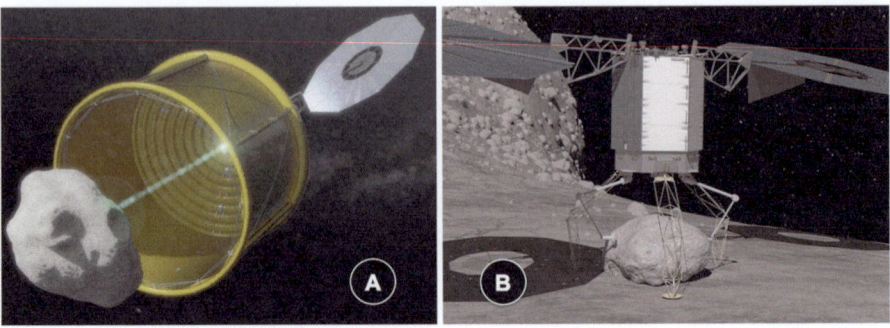

Fig. 6.3 Concepts for capturing materials associated with asteroid mining (Image courtesy of NASA.)

tant, the "territory" of space and its wealth of resources could become a zone of contention, just like a discovered lode of precious metals.

What is clearly different here is the vastness of space. It seems an impossibility to defend or control the vastness of the cosmos like you would a fixed asset such as a fort or a gold mine. Movies such as *Star Wars* has created a visualization of armed conflict among space armies that seem quite real, but in the real universe there is no "warp speed" or instant communications across the Solar System. One cannot defy the laws of physics that suggest people and things could travel well beyond the speed of light and pop from one planet or Moon or asteroid to another.

Planetary Defense Against Cosmic Hazards

The Outer Space Treaty and the Moon Treaty explicitly bans the use or deployment of weapons of mass destruction in outer space. Among its principles, it "bars states party to the treaty from placing weapons of mass destruction in the orbit of Earth, installing them on the Moon or any other celestial body, or otherwise stationing them in outer space." It also establishes as a basic principle that the use of the Moon and other celestial bodies will be for peaceful purposes and expressly prohibits their use for testing weapons of any kind, conducting military maneuvers, or establishing military bases, installations, and fortifications (Art. IV). However, the treaty does not prohibit the placement of conventional weapons in orbit. The treaty also states that the exploration of outer space shall be done to benefit all countries and shall be free for exploration and use by all the states [13].

On February 29, 2008, China and Russia went even further in proposing a draft "Treaty on Prevention of the Placement of Weapons in Outer Space and of the Threat or Use of Force against Outer Space Objects (PPWT)." This draft treaty, although considered with the United Nations, was never adopted [14].

One of the considerations in banning space weapons is a firm determination of exactly what a space weapon actually is, and perhaps just as importantly, what is excluded. For instance, what is the status of a satellite that might be used for targeting? Or a spacecraft used for surveillance to determine where to launch a future strike? Or perhaps, even more immediately, what about a satellite equipped to provide direct instructions to a drone armed with a missile to fire? Each of these capabilities is not precisely a space weapon. But certainly all of the above examples are part of a weapon system strike capability.

And the future may become even murkier in terms of defining a space weapon as new technology evolves. Is a computer virus that disables a space defensive shield system or attacks a satellite control system a space weapon? Is a laser system or directed energy beam system in space a weapons systems or does it become one at a certain power level? At what power level does it become a weapons system? As we move beyond chemical explosives to lasers, directed energy beam devises, and electromagnetic pulse devices there will be even greater difficulties in defining exactly what is meant by space weaponry.

There is also the issue of why we might need certain systems in space. It turns out that capabilities launched into space that might be considered weapon systems are also the type of technology that might be important to defending Earth against a disastrous asteroid or comet impact. Even some future electromagnetic systems that could be used as a weapon might also be employed as part of a defensive system against coronal mass ejections. This could be important in the next few years if Earth's magnetic poles reverse and our natural electromagnetic shields, i.e., the Van Allen Belts, are greatly weakened. In prohibiting the deployment of "space weapons" we could also serve to eliminate the possibility of defending our planet from a natural cosmic hazard that is far more dangerous than a human attack.

The modern world is filled with dilemmas. It is really hard to come up with space rules and regulations that protect us from each other without creating unintended consequences. As we recognize that asteroid strikes are four times more common than previously thought and Earth's natural electromagnetic shields are seriously weakening, it is becoming clear that space protective systems are something that we need to seriously consider. Indeed space protective systems may one day include new types of space systems that allow us to cope with climate change as well.

Charting a New Path Forward

The questions that now confront the world in the space arena abound. This is particularly true as new space technologies, new types of space weapons, and new space transportation systems evolve ever more rapidly. Just some of the questions that are relevant to today's world are:

- What substantive rules are needed to govern the conduct of nation states, particularly in the event of an armed conflict in space?
- What can the United Nations, its specialized agencies, or other international entities do to alleviate conflicts in space and create new regulations or standards to minimize the possibility that misunderstandings will occur?
- In particular, what new reforms in space law and regulation, oversight by international organizations, new national laws, codes of conduct, or confidence building measures, such as transparency, would help to alleviate future space conflict?
- What are the constructive and proactive roles that the International Civil Aviation Organization, the International Telecommunication Union, the Federal Aviation Administration (FAA), the European Aviation Space Agency, and other national and international agencies might contribute to the safety and security of the protozone?
- What are the most urgent and achievable forms of global space governance that could be agreed on and enacted to prevent strategic conflict in space or their ground-based control systems?

These and other questions are the new types of space dilemmas that confront us today and those who would seek to bring peace not only to our world but to the cosmos that surrounds us. Rapid technology change and the economic opportunity presented by a new space economy make this challenge even greater.

Conclusions

The current challenges associated with space security are growing in both scale and complexity. There are the new challenges associated with the rapid development of commercial entrepreneurial space systems. One of the newest elements is the deployment of a whole new range of capabilities in the so-called protozone—from spaceplanes and hypersonic transport to high altitude platform systems, robotic freighters and military surveillance systems.

There is also the increasingly large population of vital satellites in Earth orbit. The thousands of operational satellites and tens of thousands of debris elements continue to increase the potential for interference, jamming and disruption to key satellite systems. Such events threaten the operation of many essential global functions. Thus, "hostile actions" in Earth orbit against these resources could be considered an act of war. In fact a number of spacefaring countries have formally declared that they indeed would see an attack on their satellites as an act of war.

On top of these space challenges involving Earth orbit, there is the new issue of space mining and other space activities that would be carried out beyond Earth orbit. And there is the very real danger of future deployment of space weapons and various types of conflicts in space that can occur in future decades. What is clear is that there will be billions of dollars invested in strategic space systems in coming decades and that this will indeed be a part of the new gold rush.

This chapter outlines some of the more apparent areas of possible disputes in space or conflicting uses of space that could put countries or commercial enterprises at odds. Those space security concerns discussed in this chapter are just the ones that are now becoming apparent or are likely to evolve in future days. But this list is certainly not exhaustive. Indeed some of these future security issues may not come from conflicting nations or new space industries but rather from what might be called cosmic hazards. These significant hazards from space include asteroids, comets, or gigantic solar storms or even from runaway cascades of space debris that could threaten current and future space assets. These cosmic hazard issues are addressed later in this book.

All of these strategic challenges will very likely increase in intensity and complexity as space commerce and future cosmic exploits become more and successful and vital. "Security in space" has always been far from certain, and the bottom line is that risk of conflict in space is unfortunately growing as we see the maturing of anti-satellite technology, more proposed use of the protozone, expansion of critical space infrastructure, and space mining and solar power satellites.

References

1. Space Security Index 2015. http://spacesecurityindex.org/. Last accessed May 2016.
2. State of the Industry Report, 2015 by the Satellite Industry Association, The Tauri Group http://www.sia.org/wp-content/uploads/2015/06/Mktg15-SSIR-2015-FINAL-Compressed.pdflast. Accessed December 26, 2015.

3. "If there were a day without satellites", produced by Joseph N. Pelton and Ricky Benedict videographer, December 2015. https://www.youtube.com/watch?v=5sgM7YC8Zv4. Last accessed December 26, 2015.
4. https://s.yimg.com/fz/api/res/1.2/1yYi.QEXa3uhCevdGGCtDw-./YXBwaWQ9c3JjaGRkO2g9ODQ0O3E9OTU7dz0xMjI2/http://www.dmg-federal.com/wp-content/uploads/2013/03/Figure-1.jpg.
5. Joseph N. Pelton, "A New Integrated Global Regulatory Regime for Air and Space: The Needs for Safety Standards for the protozone, Manfred Lachs Conference, McGill Institute for Air and Space Law", May 2014.
6. Joseph N. Pelton and Peter Marshall, *Launching into Commercial Space*, (2015). AIAA, Reston, VA.
7. The Chicago Convention for Civil Aviation of 1945. http://www.icao.int/publications/pages/doc7300.aspx. Last accessed as of December 28, 2015.
8. Joseph N. Pelton, New Space and Protozone Transportation Services, ICAO Space Symposium, March 18-20, 2015, Montreal, Canada.
9. Ibid.
10. http://www.ainonline.com/sites/default/files/uploads/ituingenevaresize.jpg.
11. H.R. 2262 U. S. Commercial Space Launch Competitiveness Act, known as the Space Act of 201 adopted November 25, 2015. https://www.congress.gov/bill/114th-congress/house-bill/2262/text. Last accessed Dec. 28, 2015.
12. Ibid.
13. Christopher Moore, Technology Development for NASA's Asteroid Redirect Mission IAC-14-D2.8-A5.4.1. September 2014. https://www.nasa.gov/sites/default/files/files/IAC-14-D2_8-A5_4_1-Moore.pdf.
14. Treaty on Prevention of the Placement of Weapons in Outer Space and of the Threat or Use of Force against Outer Space Objects (PPWT), http://www.cfr.org/space/treaty-prevention-placement-weapons-outer-space-threat-use-force-against-outer-space-objects-ppwt/p26678. Published in 2008 and last accessed on Dec 28, 2015.

7

Protecting Earth from Space Junk, Cosmic Hazards and Climate Change

Introduction

Each advance in space technology, just as is the case with all technologies, generates new positive payoffs and capabilities. But—and unfortunately this is a big but—these innovations also ultimately give rise to a new range of concerns, dilemmas and sometimes quite vexing problems.

The London Council around 1900 thought the automobile would solve their environmental pollution and fiscal problems. At the time they were overwhelmed by the cost of removing tens of thousands of pounds of horse poop from the streets each day. In 1900 this was a back-breaking task that was literally bankrupting the city and an army of poop-removers. Today auto fumes are choking our cities. Those planning to drive into New York City or London must pay a high premium for the privilege. And thus it seems to go. Technological fixes give way to technological problems. This cycle, which turns one generation's technological innovation into the next generation's dilemma, seems to be forgotten time after time. The prime job of futurists is not to envision new technologies. Their most important function is to probe how new technologies will create future problems.

As our population grows and we become more dependent on modern infrastructure—both on the ground and in space—we become more vulnerable to such threats as orbital debris and a variety of cosmic hazards (i.e., asteroids, solar storms and even changes to Earth's protective magnetosphere). Just as we must invest in the maintenance, repair and operation of roads, bridges, dams, power grids, and utility systems, we will increasingly need to create or upgrade space systems so that they are less vulnerable to natural and human-made cosmic vulnerabilities that could put billions of people at risk. Today, if we were simply to lose the GPS Navstar system, the loss of this single space system could be almost catastrophic. Take out the GPS, and it would adversely affect electronic time stamping of banking operations, the synchronization of the Internet in most countries, the takeoff and landing of aircraft, the routing of ships and automotive traffic, the launching of national missile defenses, and maybe even speed dating.

The more technology we acquire and use, the more vulnerable we become. Modern humans have a much higher standard of living today, but one of the prices we pay is almost total dependence on electric power systems, modern transport and distribution systems, banking systems and more. The frightening question raised by this book is, could we survive the loss of our space systems?

The Increasing Risks from Cosmic Hazards

We humans today are truly more vulnerable as a post-industrial society than we were even in the days of the caveman. It is much easier to find a new cave, and to find a new place to hunt and forage, than to build a new city and a completely new modern infrastructure.

And now we learn that the level of threat to humanity from cosmic hazards is greater than previously understood. How is that possible you say? Haven't the cosmic threats always been there? For the most part the cosmic threats have not changed—although we have managed to invent a new human-created problem of space debris. However, Earth's protective magnetosphere is shifting and in effect weakening at a time when we have much more vital infrastructure to protect. Here is what we know.

A combination of space probes, ground observatories and electronic sensors deployed around the planet have revealed in recent years that the cosmic hazard of asteroids are four times larger a problem than we had thought. It turns out that 60 tons of cosmic dust falls to Earth every day, or at least 22,000 tons a year. Some estimates place it as high as 40,000 tons that settles to Earth in a year. If Earth were a person he or she would have to go on a serious diet!

The data from nuclear explosion detectors on the GPS satellites over the past decade have verified that the constant bombardment of asteroids, bolides, meteors and meteoroids has been underestimated by perhaps four to five times—both in frequency of occurrence and mass.

The issue of increasing risk from cosmic hazards is building up here on Earth. Increasing risk comes from an ever increasing human population and the rise of megacities—massive, sprawling cities with more than 10 million people. There will be more than 50 such cities by 2050 and perhaps a 100 by 2100. This means that when asteroids carry out target practice on our small planet they have more and more exciting targets to hit. We were lucky with the cosmic hit that led to the 40-m object hitting the very remote Tunguska area of Siberia in the 1900s. The next time it may be the San Francisco Bay area, which is of an equivalent size to the devastation area burned to a crisp when the rogue Tunguska object hit a century ago.

In short, the rising problem of cosmic hazards is greatly driven by changing patterns of consumption, transportation and energy—plus the greatly expanded urban area that is creeping over our planet. Just two centuries ago we humans were quite rural. Global civilization was less than 5% urban. But today we are 53% urban and headed toward being 70% urban. These shifts in population and dependency on modern infrastructure add up to much greater risks and vulnerabilities. The human population and consumption is growing, and the available natural resources such as oil and drinkable water are shrinking. This is not a formula for success.

The truth is that the accelerating needs generated by a global service economy will ultimately serve to increase the threat to humanity. If a dinosaur-killing asteroid or comet should come to visit our small planet and crushes our electronic grids and shrouds our planet in clouds of dust that block out the Sun, we will be ill-prepared.

And, there is an even bigger threat, in terms of likelihood of occurrence, and that is coronal mass ejection. Today's modern economies and patterns of employment have created a world that is increasingly dependent on access to electric power grids, application satellites and other vital infrastructures vulnerable to a solar storm like the Carrington event that occurred in 1859. At that time the Carrington telegraph offices were lit on fire and the northern lights were seen as far south as Cuba and Hawaii. A repeat of this event today with trillions of ions traveling at a reasonable fraction of the speed of light would have blown out electrical transformers, crippled satellite networks, and disabled vital infrastructure across the planet.

Earth's magnetic poles keep in place the highly charged energetic ionized particles that constitute the Van Allen Belts. These belts act as electromagnetic shields that protect us from harmful space weather.

Unfortunately, Earth's magnetic poles are not behaving themselves. They appear to have gone on vacation. The north magnetic pole has moved down to Siberia and headed south. The south magnetic pole, not to be outdone, has headed north. These excursions, as confirmed by NASA's MMS satellites and ESA's Swarm satellites, are bad news. Some computer profiles prepared in

Germany suggest that the Van Allen Belts might be reduced to 15 % of their effectiveness in warding off a major coronal mass ejection.

And asteroids, comets and coronal mass ejections are not the only hazards. Solar flares can cause genetic damage and elevate the incidence of cancer. Each year the number of derelict spacecraft and upper stage launchers left circling our planet continues to rise. Space debris continues to accumulate, with some 23,000 space objects larger than a baseball being tracked. Dr. Donald Kessler, noted for the "Kessler Syndrome" that projects a runaway cascade of space junk, has forecasted a major space collision every 10 years. This projection is based on there not being any new launches or new orbital debris. Estimates prepared by the European Space Agency for a conference on space debris suggests that Kessler estimate is too cautious. The ESA estimate is that every 5 years there will be a major orbital debris smashup that will serve to make the problem worse. If either of these assessments is right, orbital debris will accelerate in frequency over time. Thus, orbital debris is becoming a space hazard as well.

This chapter reviews the status of current cosmic hazards that face nations of the world and strategies for dealing with these challenges through better detection, a more proactive technological response, the creation of new standards, legal or regulatory actions or possibly new organizational mechanisms. Fortunately some of these steps have been initiated, but there are still many more that need to be taken.

Potentially Hazardous Asteroids and Comets

If a relatively large asteroid or comet were to strike Earth, it could wipe out a megacity or worse. There are many variables to be considered—the size and composition of the asteroid, the speed of the asteroid or comet, etc. In this case it would likely be traveling at a very accelerated pace. Finally, the nature of the impact in the atmosphere, whether on land or a water impact that could trigger a truly massive tidal wave, is also key. Such a collision by a large space rock just off shore in the ocean could kill and/or injure a considerable number of people, not to say ocean wildlife.

The nature of the impact and the recovery might be greatly different on a local, national and a global scale. Estimating the impact and the length of time of the impacts would be difficult, but a large enough space rock could be enormously difficult and financially devastating.

The so-called K-T mass extinction event some 67 million years ago destroyed at least 70 % of all plant and animal species and has been widely conjectured

7 Protecting Earth from Space Junk, Cosmic Hazards and Climate…

Fig. 7.1 Photograph from the International Space Station of the remnants of the K-T crater along the Atlantic Ocean off the coast of Mexico (Image courtesy of NASA.)

to be the trigger that wiped out the dinosaurs. Today the options are manyfold. There could be an asteroid strike similar in magnitude to the K-T mass extinction event that perhaps smashes into an ocean and creates a tidal wave that creates massive urban destruction from Boston, Massachusetts, to Miami, Florida, or perhaps an equally destructive path of terror almost impossible to imagine along the coast of Europe, the U. K. and Africa (Fig. 7.1).

However, this would be only the first step in an ongoing disaster. A thick morass of debris would rise up as a dense cloud of particles that would block out the Sun for months if not years. Crops would die worldwide, and those not killed as a consequence of the first strike might die a slow death of starvation. Such a true "black swan" catastrophic event is truly something the modern world is not prepared for in any real way.

We are not prepared in terms of prevention, warning, mitigation, or recovery. This not to say that improved tracking and warning systems could not be devised and that long-term warning systems could allow effective preventive actions to be taken many months if not years in advance.

The following discussion outlines the nature of these asteroid and comet-based cosmic threats and the types of technical and operational tracking and warning systems that could give us an improved warning system against destructive space rocks raining down from the skies.

Space agencies, observatories, and even amateur astronomers have been on the lookout for monstrous killer space rocks for some time. When the U. S. Congress and NASA first began to seriously worry about cosmic threats from potentially hazardous asteroids, the first established goal was to find all near Earth objects over 1 km in diameter. The good news is that some 90 % of the most damaging space rocks have now been located—especially via the good efforts of an infrared telescope called NEOWISE. This acronym stands for Near Earth Object Wide-field Infrared Survey Explorer [1]. In this case one should never believe the old saying that goes: "What you don't know can't hurt you." The unknown asteroids and indeed the undiscovered comets can hurt us a lot.

When the NASA scientists came to the U. S. Congress with their report on their efforts, there were a number of questions asked, and the next set of guidance was set in December 2005 with Section 321 of the NASA Appropriations Act. It specified that NASA was to go out and find those asteroids that are 140 m in size or larger. Now one might think that this new guidance was based on the fact that someone had done some very careful studies and concluded an asteroid less than 140 m would not be too dangerous and cause minimal harm. The truth is quite the opposite.

To get an idea of what size of asteroid represents a major city-killing event one only has to look at the evidence from the Tunguska event of June 30, 1904. This space object, either an asteroid or a comet, traveling at 54,000 km/h (33,500 mph) exploded some 8 km (5 miles) above the Siberian forest with the explosive force more than 1000 times that of the Hiroshima atomic bomb and flattened and incinerated a forest in a radial design some 2000 km^2 wide and containing 80 million trees.

This object, which could have wiped out San Francisco and Silicon Valley, was not 140 m in diameter, not 100 m, not 75 but rather only about 40 m in diameter. According to Dr. Donald Yeomans, head of NASA's Near Earth Objects office, this rather "insignificant space rock" was able to do an incredible amount of damage [2].

The destructive impact of an asteroid that was indeed as large as 140 m in diameter and traveling at perhaps 50,000 mph (80,000 km/h)—a more typical speed for an asteroid traveling relative to Earth orbit—would collide with a force that was some 65 times greater. If one does the math and multiplies the force of the Japanese nuclear bomb 1000 times again by 65 one gets the amazing result of the equivalent of 65,000 Hiroshima bombs. In short, there is a really important issue here—that the guidance given to NASA by the U. S. Congress for what to look for is "wrongheaded."

In fact, asteroids some 35 m in diameter should be considered major city killers. In terms of mass this is an object that is 43 times smaller than those being searched for in the NASA search protocol. And that is the least of the bad news. As one looks for smaller and smaller asteroids the numbers go up exponentially. There are likely to hundreds of times more potentially hazardous asteroids that are 35 m or more in size than there are asteroids that are 140 m in size.

NASA's efforts are far behind schedule, and they have found only a fraction of the asteroids that are over 140 m in diameter.

The estimates from the B612 Foundation, which was started by astronaut Rusty Schwieckart and now headed by astronaut Ed Lu, is that there are on the order of 1 million near Earth objects (NEOs) that are 35 m in diameter or larger. Probably all of these at the right velocity and angle of entry would be able to wipe out metropolitan New York City, Tokyo, or Shanghai. As the number of megacities continue to increase toward 50 or more, the chances of a mega-kill by a potentially hazardous asteroid continues to increase.

Let's put what we are talking about into numbers. At best NASA, the other space agencies and the ground observatories that are looking for potentially hazardous asteroids has found about 20,000. This would leave 980,000 potentially lethal potentially hazardous asteroids to be found and assessed as to the likelihood that they might wipe out Tokyo, London, or Shanghai with some 10 million people in a brief moment. In a game of cosmic Russian roulette these are lousy odds.

NASA has a pending, but unfunded, project known as NEOCam that might speed up the tracking of asteroids 140 m in diameter. The B612 Foundation has their own project known as Sentinel that is geared to find asteroids down to 35 m in size. Amazingly they have raised from private sources about half of the $400 to $500 million that is funding the project.

Rusty Schwieckart and Ed Lu have said their project is equivalent in cost to that of an Interstate highway exchange. They have pled their case to Silicon Valley billionaires with some success. Their sales pitch is along the following lines: If NASA won't make defending our planet a top priority, then help us to plot the orbits of the most dangerous space rocks and to find a much higher percentage of the city killers. Help us to build a satellite with better technology to spot dangerous space rocks and do so up to a century in advance. Currently the jury is out as to what NASA and the B612 Foundation will be able to accomplish and when. Let's pray that between the two and the other space agencies we can find better ways to spot asteroid threats.

The truth is that there are "black swan" events triggered by cosmic hazards, and the magnitude of that threat is real. A truly massive hit is not only a threat to our biggest cities but to modern society as a whole [3].

Major Solar Storms

The risk—or at least the probability—of a major solar storm hitting Earth is much higher than either an asteroid or comet. NASA has done risk assessments and concluded that there is perhaps a 15% chance of a major solar strike within the next decade.

A massive solar flare of X-rays and even more energetic UV radiation with great power can erupt at any time and elevate the risk of cancer and genetic mutation. During the 11-year-cycle between solar max and solar minimum the likelihood of damage varies greatly. Damage is indeed some 16 times more likely during solar max. The danger of solar flares is particularly elevated where the atmosphere is weakest, in the polar region and where ozone holes are found. The higher incidence of skin cancer in locations such as New Zealand and southern Australia has been unfortunately been well documented. Frogs in the highest latitudes have demonstrated genetic mutations as well [4].

Satellites above the protective shield of Earth's atmosphere can, of course, also be hit by major solar flares. Typically satellites are powered down when there are known to be high levels of radiation or ion storms coming from the Sun's corona. There have been satellite failures attributed to solar storms, but sometimes it is not clear what has caused a satellite's failure. In some cases a particular component fails, but the satellite can be restored to full or partial service by using redundant backup components on board the satellite. More heavy duty circuit breakers and better solar event warning systems can help to provide enhanced protection, but the key may be having spare satellites for critical systems stored in protected areas ready to launch quickly after a truly major hit.

The greatest danger of all comes from the ion storms that can come from the Sun in the form of coronal mass ejections (see Fig. 7.2). In the middle of the nineteenth century, the risks to electrical systems were small because such systems did not exist. The Montreal event of 1989 took out electrical systems from Chicago to Montreal, and the Halloween event of 2003 damaged the electrical grid in Scandinavia. These are real-world events that are indicators that we are today increasingly vulnerable to major coronal mass ejections. Heavy duty CMEs are more powerful than a nuclear blast by far. They can create natural electromagnetic pulses (EMPs) that can outdo the biggest of manmade bombs.

If these eruptions of ionic mass—rather than electromagnetic radiation—should burst out from the Sun in just the right direction to hit Earth, then great damage to a wide spectra of modern infrastructure is not only possible but guaranteed.

Fig. 7.2 A major coronal mass ejection could destroy electrical grids and satellite nets (Image courtesy of NASA.)

A massive enough coronal mass ejection hitting Earth could possibly damage and take offline critical satellite networks, cripple electrical grids and pipelines, and possibly adversely affect information and communications networks and defense systems. If the GPS system were to be disabled, for instance, then it is possible that we would soon lose the synchronization of the Internet in most countries, bank transaction time stamping would be lost, aircraft could be grounded and ships at sea would lose their efficient guidance systems. Chaos would reign.

And this, unfortunately, is not all the bad news. There is new data from the ESA Swarm satellite probes just launched in 2015 as well as NASA's MMS satellites that confirm that Earth's magnetic poles are shifting. This movement, with the north magnetic pole now moving as far south as Siberia, suggests that the protective shield represented by the Van Allen Belts—and controlled by the geo-magnetosphere—are weakening. In short our national shields against a massive CME storm are coming down. Computer modeling suggests that the natural shielding may eventually be down to 15 % of what the Van Allen Belts provide to us today [5].

There is now coming into focus a rather disturbing equation. This equation shows on one hand an increasing global population, more and more electrical grids, more electrical, computer, and remote control devices and satellites, and more and more dependency on modern infrastructure. On the other hand we have a global atmosphere with a weakening protective ozone layer due to pollution, a geo-magnetosphere with shifting magnetic north and south poles, and a Van Allen Belt system that is less able to protect our vital infrastructure against solar storms. This equation seems to spell out the words—big time trouble!

The Space Debris Problem

If these cosmic worries were not enough we have one more thing to worry about in the skies. The development of space debris and its increase over time is a story that can best be told visually. In 1960 there was no space debris. By the 1980s scientists such as Donald Kessler was noting the many acts that were contributing to the build-up of space junk and warned that if steps were not taken to mitigate this trend then over time there could be a runaway cascade of space debris that would create a deadly build-up. Chain reaction collisions would then create a deadly cloud of debris elements that could eventually deny safe access to space.

At this time natural debris in the form of space dust and meteoroids constituted the far larger problem, and modest efforts were made to control the space debris problem. Thus explosive bolts were no longer used, and efforts were made to de-gas fuel tanks that might explode. Today the problem has become progressively worse. The Chinese hitting a defunct Fengyun IC with a missile in January 2007 created 2317 pieces of trackable-size debris elements and an estimated 150,000 debris particles. NASA calculations suggest that even by 2035 30% of the debris larger than a baseball would still be in orbit [6]. This was followed in February 2009 by the accidental collision of the Russian Kosmos 2251, a defunct military communications satellite, with an active Iridium satellite to create on the order of 2000 new debris elements [7] (Fig. 7.3).

There are now recommended procedures in place to seek to reduce space debris, but actual practice in terms of more and more satellites and especially the launch of large-scale constellations of satellites in low Earth orbit strongly suggests that the orbital debris problem under current activities will just get worse. The French law that puts enforcement power behind the requirement for all launched satellites to be de-orbited 25 years after end of life is currently one of the most important measures seeking to enforce debris mitigation. Plans by OneWeb to deploy some 800 low Earth orbit satellites and rumors that SpaceX might deploy a 4000 low Earth orbit constellation have occasioned even greater concern recently that the Kessler Syndrome could materialize sooner rather than later.

Fig. 7.3 A representation of current space debris in low Earth orbit (Image courtesy of NASA.)

You might ask at this point, well what should be done? Although the agenda is far from clear, action is clearly needed. We should cooperate together, especially among the space agencies of spacefaring nations, to achieve a truly effective planetary defense program against comets, asteroids, solar storms, and even man-made threats such as orbital debris. This may take decades to put in place, but efforts should be re-doubled now. The problem is that public opinion has only a modest understanding of the nature and size of the threat. And space agency leaders around the world see expenditures for space defense as cutting into their own space development and research programs. Necessary programs need to go beyond NEO tracking and shielding. We need to include serious efforts to address and protect against massive solar storms as well as coming changes to the magnetosphere. These new planetary defense programs should also aim to take on orbital space debris and even climate change. In truth protection against severe solar storms and related technological systems to guard against significant climate change could turn out to be potentially closely interlinked.

These concerns as well as a possible planetary defense program, will be addressed in summary form below.

Action Plan for Planetary Defense Against Cosmic Hazards

Here are some of the key steps that might be taken:

- Doing national and international risk assessments of the nature of cosmic risks from asteroids, comets, solar flares, coronal mass ejections, magnetosphere shifts and reduced protective shielding from the Van Allen Belts, the risks associated with orbital debris increases, and also identification of possible protective strategies against all types of cosmic hazards and whether there are proactive strategies that might be pursued to reduce these risks. This assessment would also include consideration of whether there are new commercial or other space industries that might be created as part of a planetary defense capability. Shield systems to cope with severe solar storms might be alternatively used to generate solar power or to moderate the degree of climate change. Most exotically of all these systems might be used to allow the creation of a breathable atmosphere on Mars.
- Moving to place planetary defense as a top priority and even a primary goal for space agencies. This would involve consideration of specific steps such as creating new senior administrative positions within a space agency to undertake solar research, understand changes to Earth's magnetosphere and also related Van Allen Belt research. It would also involve increasing asteroid and comet tracking capabilities, develop new technologies to deter or redirect asteroids and comets, research and create orbital debris removal capabilities and other related technologies and systems. These new positions in the space agencies would also be empowered to engage in international consultations to develop a global agenda for cooperative programs in these areas. After all, we are all in this together.
- Pursuing innovative programs to identify and track orbits of potentially hazards asteroids and comets, to identify and better understand the nature of solar storms and changes to the magnetosphere, and to develop creative "outside the box" systems to defend against cosmic hazards.
- Supporting the U. N. International Asteroid Warning Network (IAWN) and the Space Planning Mission Advisory Group (SPMAG) and its expanded capabilities [8].
- Expanding efforts to reduce orbital space debris through active debris removal with innovative systems that include new ideas such as systems that operate using Earth's magnetic field, or new systems that capable of de-orbiting multiple pieces of space debris.

References

1. Wide-range Infrared Survey Explorer, http://www.jpl.nasa.gov/wise/. Last accessed Dec. 29, 2015.
2. The Tunguska Impact, *NASA Science News*, June 2008. http://science.nasa.gov/science-news/science-at-nasa/2008/30jun_tunguska/. Last accessed Dec. 29, 2015.
3. Pelton and Allahdadi, *Handbook of Cosmic Hazards and Planetary Defense*, Editors' Introduction, 2015, Springer Press, New York.
4. Ozone and UV: Where are we Now? http://www.skincancer.org/prevention/uva-and-uvb/ozone-and-uv-where-are-we-now. Last accessed on Dec. 30, 2015.
5. ESA's Magnetic Field Mission: http://www.esa.int/Our_Activities/Observing_the_Earth/The_Living_Planet_Programme/Earth_Explorers/Swarm/ESA_s_magnetic_field_mission_Swarm. Last accessed on Dec. 30, 2015.
6. Zachery Keck, China conducted anti-satellite missile test, *The Diplomat*, July 2014. http://thediplomat.com/2014/07/china-conducted-anti-satellite-missile-test/Last. Accessed Dec. 29, 2015
7. Becky Iannotta and Tariq Malik: "2009 U.S. Satellite Destroyed in Space Collision" *Space.com*, Feb. 11, 2009. www.space.com/5542-satellite-destroyed-space-collision.htmlLast. Accessed Dec. 29, 2015.
8. Space Mission Planning Advisory Group: http://www.cosmos.esa.int/web/smpag. Last accessed Dec. 29, 2015.

8

Space Habitats, Space Colonies and the New Space Economy

Introduction

Most of our visual images of space habitats and space colonies come from science fiction movies. Perhaps you have seen movie classics that range from the sexy and garish worlds portrayed to us in *Barbarella, A Clockwork Orange,* and *Starship Troopers* to the more ethereal and philosophically challenging futures that Arthur C. Clarke and Stanley Kubrick visualized for us in *2001: A Space Odyssey*. These imagined alien worlds were, in quite different ways, all strangely visually compelling and often frightening. But essentially all these worlds were also far removed from a real and actually possible future that could ever exist. The most realistic representation that has recently been provided to us on the big screen is *The Martian*. This movie at least conveys the magnitude of the struggle to stay alive in a stark and cold Mars environment.

If we want to understand the challenges of creating future space habitats and off-world colonies, it is probably much more useful to go to the labs of space agencies and research institutes or to visit the New Space enterprises that are intent on making the future happen as quickly as possible.

Today there are real world companies that are not trying to imagine a future world in space but rather trying to invent it. Enterprises such as Deep Space Industries, Planetary Resources, SpaceX or Bigelow Aerospace are actually intent on developing the technology to create a new off world reality and invent space habitats in which people can live for extended periods of time.

When Peter Diamandis said: "The meek shall inherit the world. The rest of us will go to Mars," it was less of a joke and more of a manifesto. Few entrepreneurs in the space industry have been so intent on trying to realize a new and viable space future where people could actually live and exist on other celestial bodies for prolonged periods of time. Part of this intensity was to make the new space future happen within his lifetime. But Peter Diamandis and Elon Musk are far from alone. There are now thousands at work to create spaceplanes, design space habitats, and to develop hundreds of new space-related technologies that are crucial to off-world enterprises.

In 1972, on "The Voyage Beyond Apollo" aboard the SS *Statendam* ship, Neil Ruzik gave a talk in which he described the dozens of his patented inventions that could only function on the Moon or in outer space. This author was aboard that ship. It was then that the real challenges that would be presented if people were ever to actually live, function and even reproduce in outer space became clear. It seemed then as we discussed the future of space on board the ship that carried us down to the Kennedy Space Center to watch the Apollo 17 launch that space colonies actually might be possible in the span of less than a century—perhaps much sooner. Today, a half century later, the likelihood of viable space colonies certainly seems much further than 50 years in our future.

When Elon Musk explains in some detail his thoughts about how to create a functioning and completely self-sufficient colony on Mars that would entail the ability to sustain 1 million persons, it seems unlikely that Musk will see his dream realized during his lifetime. Manned missions to Mars within 50 years, perhaps yes. Sustainable habitats on the Mars and on the Moon will likely come later. But if humankind survives it will create permanent presences off Planet Earth. This is in our DNA.

Yet sustainable colonies that can pull off self-sufficiency and long-term viability, this could be a century or more away. The question thus becomes not whether space colonies are possible. The question is how long and at what cost could such off-world ventures become sustainable? A "human mission" to Mars is one thing. A million people living, growing crops in hydroponic greenhouses and sustaining themselves on the Red Planet is quite another. It took about 4.5 million years for the Southern Ape Man to get to today's modern world on Planet Earth. It certainly might take a thousand years to get to something like Elon Musk's vision of a million person sustainable space colony established on Mars. Humans very well may have to prove they can make Earth sustainable for the long term, before we can undertake mega projects as ambitious as a terraforming operation on Mars that could support human life for the longer term future. Key space-based construction efforts to create a massive shield to protect a Martian atmosphere from solar wind may need to be a prelude to something like a million person sustainable colony on Mars.

The effort to create a complete and successful off-world economy that is fully self-contained, sustainable and economically viable was not just an idle speculation by Musk. Rather it was a holistic vision of a functioning ecosystem that was complete. In other words he was not seeing Mars as a research mission. Rather his vision was that of a completely integrated and terra-formed habitat geared to independent existence for the longer term.

In short Musk was envisioning an entirely new "off-world" world, where all aspects of contemporary life would exist. Thus the new Mars would include agriculture, mining, material processing and manufacturing, banking, transportation, communications, electrical power grids, water, sewage and other utilities. It would be a world capable of building construction, complete educational and health care capability and all other aspects of life that include bars, dance halls, cafes, restaurants, hotels, nightclubs, movie theaters, and strip clubs. Presumably there would also be a police force, courts, unemployment benefits—perhaps even crime and racial and ethnic intolerance. Human technology and technical capabilities continue to grow, but it seems this is not so with our temperament and our social and cultural perfection. The Russian space pioneer Konstantin Tsiolkovsky said that human space exploration would "perfect the human soul." Time will tell.

Elon Musk has thought centuries ahead and envisioned that the threshold for a complete off world society with all the beauty, innovation, creatively and "warts" of human civilization could 1 day be present on a Martian colony—a colony that could grow into a mega complex. What was not clear was, of that million, how many would be humans and how many would be "thinking robots" or androids. Would a million "people" be needed for technical reasons? Would a million people be needed to create a stable and viable social milieu? Or to provide the needed genetic and thought diversity? Or yet for some other reasons?

But despite what this ultimate off-world reality might ultimately be, there are a growing number of entrepreneurs and scientists and engineers that are seeking to create space habitats, small research colonies for the Moon, for Mars, and possibly for elsewhere. These efforts by space architects go well beyond the imaginings of just a handful of space enthusiasts.

There is actually a growing membership of New Space industry entrepreneurs. These enthusiasts are at the various New Space conferences, and they actively support the efforts of the Commercial Spaceflight Federation and others that look beyond NASA and the space agencies to imagine what the future of humans in space might actually entail.

In short, there is whole new world of space entrepreneurs that are busily creating space transportation systems, off-world energy generation systems, space habitats, and more. Some are intent on developing low cost hypersonic spaceplanes as businesses here on Earth. Others are seeking new and better commercial launch vehicles that can go to Mars that rely on new propulsion systems, and yet others are researching exotic technology such as space tethers and space elevators.

Some of the areas that are getting a good deal of attention are those engaged in creating orbital habitats for space tourists and that could ultimately even lead to space hotels and long life commercial orbital research labs. Others are focused on habitats, and dwelling units for miners in space, although today's space mining companies are thinking that their operations would be conducted by smart robotic machines, 3D printers and automated material processing units with nary a human astronaut in site.

Actual corporations involved in such enterprises are designing the construction and deployment of solar power satellites, the design and operation of robotic repair and refueling systems in space, and a number of other enterprises dependent on all remote robotic operations. These are serious space business people and not movie makers with just a vivid imagination seeking to give space cadets a vicarious thrill. Some of these space enthusiasts, however, bridge the gap. Movie director James Cameron is actually a backer of Planetary Resources Inc.

Most of these visionary commercial space industrialists and engineers, though, are pragmatically seeking to design systems that are cost effective and viable components for off-world enterprises. Indeed a number of these space entrepreneurs are actually talking about projects and undertakings that can provide a return on investment.

Dr. Bradley C. Edwards, who has spent two decades at Rice University developing nano-tube carbon-based cables and researching ways to build a space elevator to geo orbit, made the leap to become a space entrepreneur. But to do this he also had to focus on how to make a living. Thus while he has scientists and engineers trying to develop ultra-high tensile strength cable to the skies, he has used this technology to design state of the art tennis rackets and sporting equipment. Many of the people devoted to future space travel have founded companies that involve various types of space tourism activities, including high altitude dirigibles. Some of the space dreamers have simply extended their businesses from today into tomorrow.

The classic case in point here is none other than the billionaire owner of Budget Suites Robert Bigelow, who founded Bigelow Aerospace to design and build space habitats. Bigelow wisely opted to start by licensing technology first developed by SpaceHab under contract from NASA. In a remarkably short period of time he has developed and launched the Genesis capsules.

These are semi-rigid inflatables that create relatively large volume habitats that are more energy and mass efficient than the metallic structures used in the International Space Station and certainly quite a bit less expensive. And hotel magnate and space enthusiast Robert Bigelow has certainly made his money back on the licensing agreement he made with NASA.

In 2012 NASA awarded a $17.8 million contract to Bigelow Aerospace to produce what is now called the Bigelow Expandable Activity Module (BEAM) space habitat [1]. This module was scheduled to arrive at the space station in 2015 for a 2-year technology demonstration. The module was completed and ready for launch on time, for the eighth SpaceX cargo resupply mission to the station, just as contracted with NASA. This launch, however, was delayed due to the July 20, 2015, launch failure of the Falcon 9.

The BEAM unit measures just 2.4 m (8 ft) wide in its packed configuration aboard SpaceX's robotic Dragon resupply spacecraft. Once it is deployed and inflated, it will add an additional 565 ft^3 (16 m^3) of volume—about the size of a double-wide house trailer. The BEAM is to be accessible to astronauts aboard the orbiting laboratory [2].

After the module is berthed to the ISS's Tranquility node, the plan was for the station crew to activate a pressurization system to expand the structure to its full size using air stored within the packed module to four times its deployed size. On May 26, 2016, however, the initial attempt to deploy the BEAM habitat structure failed to operate as planned, but on May 29, 2016, on the second attempt, the BEAM structure fully inflated to full size.

Astronauts periodically will enter the module to gather performance data and perform inspections. Following the test period, the module will be jettisoned from the station, burning up on re-entry.

The B330 (previously known as the Nautilus space complex module) is an expandable space habitat to be privately manufactured by Bigelow Aerospace. The design has evolved over time from NASA's and SpaceHab's original TransHab habitat concept. B330 will have a rather amazingly huge 330 m^3 (12,000 ft^3) of internal space. Its 330 m^3 of volume and its design by Bigelow Aerospace combine to give B330 its name. The craft will support near zero-gravity research, including scientific missions and manufacturing processes. Beyond its industrial and scientific purposes, however, it has potential as a destination for space tourism and a craft for missions destined for the Moon and Mars.

This space habitat, if deployed in Earth orbit, would have an inhabitable pressurized volume of 330 m^3 but only have a total mass of 20 tons. This compares to the 106 m^3 of volume represented by the 15-ton Destiny module on the current International Space Station [3] (Fig. 8.1).

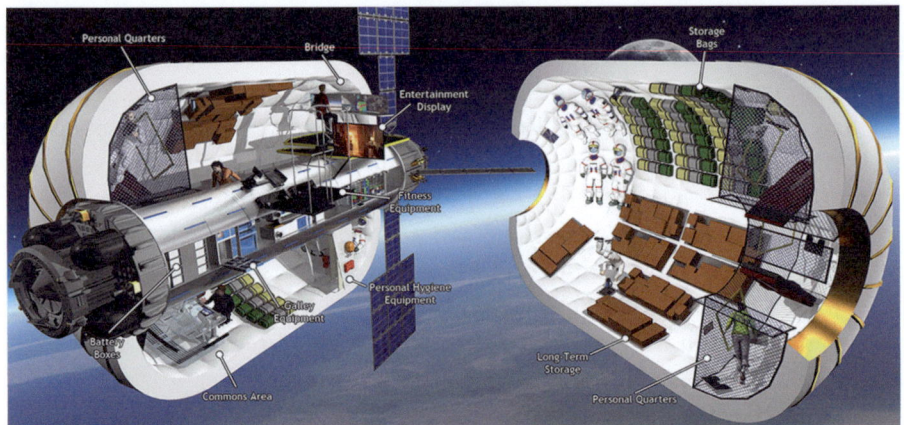

Fig. 8.1 Exploded view of envisioned B 330 space habitat with three times the habitable space as the International Space Station Destiny module (Image courtesy of Bigelow Aerospace.)

This means the B330 would provide 210 % more habitable space but with an increase of only 33 % in mass of that over the Destiny module on the ISS today. Perhaps even more significantly is the fact that because this unit would be inflated by compressed gas, it would be much easier to launch to orbit, with a lower launch cost than that required for Destiny [4].

Currently Bigelow Aerospace represents the leader in the design and construction of economic habitats that can sustain NASA astronauts, commercial astronauts and even space tourism flights or space researchers flying on behalf of commercial organizations or national agencies seeking to carry out experiments in orbit. Bigelow has already successfully flown and deployed its Genesis inflatable space habitats to demonstrate the viability of the technology. These tests have convincingly shown that the materials used in these units blown up with air are quite survivable in space. The concept explored by Bigelow Aerospace to create very large space habitats involve modular units that can be flexibly joined together (see Fig. 8.2).

Of course a sizable enough piece of space debris flying at sufficient speed to smash into the ISS or Bigelow inflatable craft could doom either of these ultimately vulnerable spacecraft. The movie *Gravity* starring George Clooney and Sandra Bullock was in many ways science fiction, but in terms of space debris the objects would move with even greater velocity and destructive force should they indeed make a direct hit. Although there is a great deal of space out there, space scientist Donald Kessler, famous for predicting the Kessler Syndrome, has suggested that every 10 years—or less—we will have a major

Fig. 8.2 Bigelow private space habitat as currently conceived with Boeing CST 100 capsule approaching to dock (Image courtesy of Boeing.)

space collision. Such an event will unfortunately generate perhaps 2000 new debris elements and the cascading debris problem [5].

The race is on, the race to launch more and more things such as space habitats and mega-low Earth orbit constellations with hundreds if not thousands of satellites in them. All of these large space objects or very numerous space objects are, of course, vulnerable to space debris. Another race is to find cost-effective and reliable ways to mitigate, reduce or eliminate the space debris problem. Some fear that the space debris buildup is winning the race toward an unsustainable near-Earth space environment. The progress in on-orbit servicing technology and electromagnetic propulsion, however, holds out hope.

The Effort to Create a Lower Cost Short-Term Space Tourism Habitat

In contrast to the sophisticated approaches of NASA and Bigelow Aerospace to create low Earth orbit habitats that might transition into a suitable habitat on the Moon and Mars, there are other, lower cost and more basic approaches being pursued by the New Space startups in the context of an affordable space tourism stay in space.

The initiative of Inter-orbital Systems (IOS) represents one such initiative. IOS of Mojave, California, is seeking to perfect a low-cost propulsion system that can be combined in multiples to create successively larger launch capabilities. This common propulsion module (CPM) is designed to create the so-called Neptune 30 that is designed to deliver 30 kg to low Earth orbit. The Neptune 45 would deliver 45 kg to low Earth orbit, and the Neptune 1000 would provide a metric ton to orbit.

The most ambitious of IOS's initiatives would be to create a one and a half stage to orbit launch system with the half stage not only serving as a propulsive unit but then serving as a temporary space habitat for tourists. The CPM currently uses white fuming nitric acid (WFNA) as the oxidizer and turpentine or furfuryl alcohol as the fuel. Although the passenger unit has been developed, IOS has quite a ways to go to demonstrate a safe and reliable high-capacity launcher system. To date they have only successfully demonstrated sounding rocket capability and are currently testing the Neptune 30. If and when IOS does develop a reliable launcher capability its space tourist temporary habitat would clearly be the lowest in cost and most basic approach to an integrated launcher and habitat system. Such a capability, however, is clearly at least 5–10 years into the future [6].

NASA and Long-Term Space Habitats

NASA in its 2016 appropriations received $55 million in new funding from Congress to develop living quarters for future missions into deep space. The U. S. 2016 government appropriations bill explicitly supports the space agency's efforts to develop a "habitation augmentation module." NASA has been instructed by Congress to proceed quickly to develop a more comfortable living and longer term facility for astronauts who will embark on long journeys into deep space. The specific objective is to develop a facility for Mars, where NASA is planning to go in the 2030s. The report accompanying the spending bill states "NASA shall develop a prototype deep space habitation module within the advanced exploration systems program no later than 2018." [7]

There have been private initiatives that might be developing a capability to go to Mars—including even one way missions—but none of these are to date sufficiently advanced to be seriously evaluated from a scientific and engineering perspective.

Moon, Mars, and LaGrange Point Habitats

Currently NASA has several projects under contract to begin the development of such an expanded astronaut habitat. The work by Bigelow Aerospace on the BEAM and B330 projects are clearly indicative of this type of development work. Some of the concepts being explored, however, are more advanced in concept and much more oriented toward exploring how to establish a long-term space colony on a much larger scale and through the use of off-world resources. One such exploratory project is with the Pacific International Space Center for Exploration Systems (PISCES),

This PISCES-NASA project involves 3D printing on a construction scale using basalt. Here on Earth, 3-D printers in prototype form are already building houses using recycled materials. Launching construction materials for a space colony via a rocket is much too costly. Basalt rocks, which are abundant on Earth and on many celestial bodies including the Moon and Mars, could be the key to building infrastructure in space using robots. Because Hawaii's basalt is quite similar to the regolith found on Mars and the Moon, it could be used to 3D print shelters, landing pads, and tools that could be used on other planets. If we could employ the building materials available on Mars or the Moon then construction costs become much cheaper.

Due to the exciting potential of such technology, Pacific International Space Center for Exploration Systems (PISCES) is one of four partners chosen by NASA to work on a project involving robotic-enabled construction. Hawaiian lawmakers—through Hawaii Senate Concurrent Resolution No. 83—have backed this initiative with local funding support. Under the proposed plans, PISCES and NASA would work together to 3D print a landing pad, a curved wall and a dome-shaped structure in Hawaii using basalt. The first phase of the project began in October 2014 and is expected to continue through 2017 [8].

Although this may sound exotic there are already large scale 3D printers large enough to "print out" an entire house in a single day. Figure 8.3 below shows a scale model of a 3D printer designed by Prof. Behrokh Khoshnevis of CRAFT, which is a research center at the University of Southern California (USC) [9].

Gerard K. O'Neill Space Colonies Concepts in "The High Frontier"

Some concepts for space colonies that have been conceived envision them not on the Moon or Mars but actually in space itself. Gerard K. O'Neill thought that it would be possible to create giant space colonies that might perhaps be deployed in a LaGrange point and that this might actually be a fully self-sustainable small world in space. The idea of an entirely artificial space colony in space may be an exotic concept, but it is really not entirely original to O'Neill.

Fig. 8.3 This is a scale-model of a 3D printer that can construct a concrete house in a day (Image courtesy of the CRAFT Research Center, University of Southern California.)

The idea of space travel and living in space has been in literature for centuries. H. G. Wells, Jules Verne, Edward Everett Hale, and others have written of such futures. But scientific speculation and the beginning of serious engineering texts are much more recent. Konstantin Eduardovich Tsiolkovsky (1857–1935), of Polish and Russian descent, pioneered the first serious studies of modern rocket propulsion along with the German Hermann Oberth and the American Robert H. Goddard. He is considered not only to be one of the founding fathers of rocketry and astronautics but an advocate of human space travel and the creation of space colonies that he believed would lead to "perfecting the human race." The first documented person that wrote about the geosynchronous orbit was not Arthur C. Clarke, who said this was the perfect orbit for communications satellites, but the Eastern European writer Potecnik, who wrote under the name of Hermann Nordung. Nordung envisioned giant cylinders rotating in geo orbit as the first space colonies.

Regardless of Gerard O'Neill's inspiration, he thought about how a space cylinder habitat might be created and not only about the needed transport and building materials but also about hydroponics for growing food for people and even livestock.

In the case of O'Neill's vision he saw the concept of a so-called "mass driver" device on the Moon that could provide a steady stream of building materials sent up from the surface as providing the essential enabling capability.

He calculated that the low gravity of the Moon was not difficult to overcome with a magnetic accelerator that could take materials mined from the Moon and send them to a "catcher system" that could then combine the materials to create his giant cylinders that were large enough to create a viable artificial gravity and to sustain a large population.

O'Neill's idea was to create two cylinders would rotate in opposite directions in order to cancel out any gyroscopic effects that would otherwise make it difficult to keep them aimed toward the Sun. Each would be 8 km (5 miles) in diameter and 32 km (20 miles) long. Each of the cylinders would be connected at each end by a rod via a bearing system. The rotation would provide artificial gravity via centrifugal force on their inner surfaces. Clearly such massive structures in space would only be possible to build some centuries in the future. More modest ideas include the building of a "space shield" at LaGrange Point 1 to protect Earth from violent coronal mass ejections from the Sun. Such shields would be for the coming period in the next few decades when the natural protection from the Van Allen Belts decreases to a significant degree. Such large-scale constructions in space could examine the viability of much more visionary ideas.

The vision was quite grand and also artistic in concept. O'Neill envisioned a world in space that was contoured with hills and valleys and lakes, an "Earth-like world" that could sustain a rather large population and theoretically be self-sufficient.

Biosphere 2

The idea of creating a sustainable space habitat has been envisioned by many but accomplished by few. The "biosphere" that was created in the desert of Arizona in the late 1980s, called Biosphere 2, was an attempt to create a viable self-contained world with "biospherians" inside this large structure. These would be future space colonists who were urged to grow their own crops and livestock and maintain a sustainable community.

Although careful studies were done beforehand the dozen and a half biospherians entered the sealed community to disastrous results. Despite their best efforts, it was necessary after a few months to vent carbon dioxide out of the vast greenhouse or the people inside would have died. The experiment failed in another way as well. The biospherians separated into two camps and grew quite hostile to one another. The psychological part of the experiment thus also failed.

The truth about designing a sustainable space colony is that it involves not only installing exercise equipment, a galley for storage of food and drink, or a system to recycle human waste. The greatest challenge of creating a viable space habitat or even larger space colony is to create and implement a design that keeps all of myriad variables in balance and a full range of life support systems always intact. This means always being able to sustain life support systems, oxygen supply, continuous uninterrupted power, and much more.

It is a true challenge to anticipate all the things that can go wrong and to protect against a major crisis that is much larger than that which was foreseen. Prime in this regard is to recognize that the space environment is indeed quite harsh and unforgiving. The movie *The Martian* provides an epic story of a single astronaut, actor Matt Damon, surviving for years on Mars. In truth the actual odds would have been less than one in a billion.

The loss of an oxygen supply or a hole in the habitat, a fire that destroys life support, sudden exposure to temperature extremes, and the ravages of solar storms are at the top of the list of cosmic hazards. Other concerns include such aspects as potentially hazardous asteroids and comets and extreme cosmic rays from across the far reaches of space. In the case of Earth orbiting habitats there is also the serious issue of orbital debris. The International Space Station has had to be maneuvered several times to avoid space debris; there are significant debris elements in its orbit, including debris from the Chinese missile hit on the defunct Fengyun weather satellite that occurred in 2007. Radiation, severe temperature extremes, and loss of life support systems are key hazards in space, with no allowances for even brief outages in vital systems. This is why robots are really much better to do tasks in space. They can be engineered to survive temperature extremes, do not have to eat or breathe, and can survive exposure to high levels of radiation as well as reasonable levels of ion bombardment. Smart robots are much better than humans at survivability. Their housing and power needs are much easier to accommodate.

Space Colonies Built to Survive Major Solar Storms, Extreme Temperatures and More

The most survivable early space colonies, built for longer term habitation, would likely involve using the soil of the Moon or Mars to provide protection against hazards from the Sun, such as ionic blasts from coronal mass ejections or X-ray or gamma radiation. Such underground vaults can be better protected against loss of oxygen and major breaches that could create exposure to severe cold or heat. Yet people are generally not well adapted to living below ground.

The future may thus seek new technological solutions. It may be that problems with Earth's own protective systems that depend on Earth's magnetosphere and a shift of the magnetic poles might lead to the development of space-based magnetic systems that can provide protection against violent solar storms. These might be needed to protect Earth against solar radiation and ionic blasts. These new space systems may also 1 day be needed to cope with climate change. Such technology developed as protective systems for our planet could be applied on a smaller scale and might be deployed to protect colonies on the Moon or Mars. Who knows? One day conservatives that deny climate change may decide to sign on to defending against solar overheating if major aerospace companies suddenly stand to gain billion dollar contracts to build new space shielding systems.

The earliest space colonies would most logically be designed to use automated machinery, processors, and smart robots to carry out most if not all functions. According to experts on machine intelligence and heuristic algorithms such as Ray Kurzweil the technology for automated space colonies are almost with us today. He projects that thinking robots with the intelligence of humans will be available by the 2030s. The first space colonies will likely take such extraordinary new capabilities into account in the planning process.

There is yet another aspect to the planning for future space colonies that go beyond the issue of durable machines, and instead consider the vulnerabilities of humans in isolated communities and very long duration missions. The Biosphere 2 experiments not only exposed problems with the sustainability of the atmosphere inside the biosphere but also with human governance and conflict even in small groups left to themselves for extended periods of time. Within the Biosphere 2 before the experiment was shut down the biospherians evolved into what amounted to competing teams and sharp rivalries despite careful vetting of those chosen for the project.

Other space agency projects to simulate long duration missions with test "astronaut teams" have exposed personality conflicts and even hostilities. All of these factors have to be taken into account. The list is long and daunting. Environmental protection; safety from cosmic hazards; communications; constantly available power system; logistics for food; sustainability against temperature gradients; loss of oxygen; human waste recycling; and an effective governance system—all of these demanding requirements suggest that after we accomplish human travel to Mars the first long term colony should most likely be established on the Moon, where transportation, communications and resupply is much, much easier than on Mars.

Terraforming Mars: The Moral Implications

The tremendous heat and atmospheric pressures of Venus do not make it a candidate for terraforming. The Moon has no atmosphere, and thus the focus on a possible long-term terraforming operation has almost exclusively been on Mars. There are a host of technical, operational and logistical questions about the feasibility and financial viability of being able to transform Mars so that people could safely live there.

However, beyond the technical issues, there is the moral issue as well. As noted in the first chapter, there are space ethicists who argue that such an enterprise is essential to sustain Homo sapiens as at least a two-planet civilization, as the first step to a longer term future where humans ultimately travel to the stars. There are other ethicists that argue just the reverse. They argue that humans have used up the natural resources of Earth and polluted its oceans and atmosphere. Now we are polluting Earth's orbital space with debris and creating holes in the ozone layer of the stratosphere.

Some space ethicists take a very discouraging view of humanity and its potential to do positive things. These skeptics suggest that humans are almost like a virus. They see humans as virtually infecting all the environments into which they are introduced. These skeptics suggest that if Homo sapiens move into deep space they will destroy these habitats as well. They argue that humans should learn to walk before they run. They strongly resist the "arrogance of attempting to terraform Mars until we can prove we can sustain our species on Earth and cope with climate change and orbital debris first.

Certainly it makes sense for humans to give such objectives top priority. It makes sense that we first preserve the delicate balance of Earth's biosphere, protect ourselves against cosmic hazards and also clean up space debris to make Earth orbit sustainable for the long term. If we could work together to protect Earth we might learn that this is more productive than warfare. At least it's a thought—humans working together.

Population Control Within the Constraints of a Closed Environment

One of the difficult questions that NASA and space agencies love to avoid is the question of sex and human propagation in space. Also there are enormous challenges of health care in a space environment, where doctors and nurse and hospitals will be in short supply and the chemistry of peoples and animals in space we know is different than on Earth's surface. The off-world habitats will need to be expanded and have additional facilities if people have babies in space. One thing that is clear, however. People are smart enough to find ways to have sex without gravity and even find a way for sperm to fertilize an egg even in the inhospitable theater of near zero g. It is perhaps useful to see the need for zero population growth in space to help us understand why the same is true on Planet Earth.

Our Earthly spaceship is much larger, but it too has limits as to the size of human civilization it can ultimately support. As we grow from 7.5 billion toward 12 billion that constraint will become ever clearer.

Growing New Infrastructure and Housing with Genetic Seeds

Professor Freeman Dyson at Princeton University has developed in some detail an even more exotic idea. He has outlined how that we could build colonies in space in the most efficient way possible. He has suggested that instead of building habitats in space that one could develop specialized "seeds" that could be launched to another planet. There these genetically designed "seeds" could grow into building structures. Such biological structures could, of course, be a very efficient way of transporting structures to space.

When we see limits to what humanity can do and achieve in future years it is because we do look to the future through the rearview mirror of history rather than through the telescope of tomorrow. It might be possible to design organisms that chew up regolith material and create materials that can be loaded into 3D printers and then spew out Moon buggies, tools, instruments, buildings and even robots that can build mass drivers or rockets. The reason that Ray Kurzweil called the future breakthrough the Singularity is because he foresees not a leap forward in technological progress but a true explosion in innovation. If we can truly design seeds that grow future habitats, then the timetable for off-world colonies can indeed be realized much, much sooner. Every time one discounts the possibility of rapid change just think of how in just 14 years we went from a time when there were not even any satellites to a man walking on the Moon.

Population Control Within Constraints of a Closed Environment

Too few people get the very real proposition that we travel through space on a spaceship. It travels around a Sun, and the armor of our spaceship is just a shroud of atmosphere and the Van Allen Belts. The resources that we have are fortunately renewable because our Sun provides us free energy and warmth. The carrying capacity of our spaceship is finite. We cannot keep on expanding our population forever.

In a short amount of cosmic time, we will reach the limits of our growth and then we must find a way to live within our means or new resources to fuel our growth. Some think that space mining is a key to expanded growth and long-term survival. The long-term key to survival, however, must be sustainability. Space mining without sustainability would be a modest band aid applied to the resource needs of a multi-billion-person world. We might be able to find a way to live with a world containing as many as 12 billion people in 2100, but what about 24 billion in 2200 or 48 billion in 2300? In the longer term exponential growth will kill us unless we find a way to make zero population growth work. A space colony with limited means will actually be a key experiment for us to tryout. It is crucial for us to see how humans cope with a world that finds the means to limit population growth and make such a system work.

Conclusion

Of all our aspirations for the future, sustainability should be the one thing we strive for above all others. Sustainability is the key to our future prosperity, to quality health care and education, to housing, to reasonable employment and indeed prosperity.

And what does the future hold? Prediction is always very difficult. Fast-moving technology will overtake your forecast of 5 or 10 years ahead. Ten to 20 year predictions will also be thwarted because of regulatory constraints and legislative inertia. For 30–50 year predictions most of the people that might check up on you will not be around to check on them, so these are easily the safest.

Arthur C. Clarke's three laws are as useful a guide to the future as any. They are not only entertaining but also incredibly insightful and also quite succinct. These three laws are as follows:

- *First law:* When a distinguished but elderly scientist states that something is possible, he is almost certainly right. When he states that something is impossible, he is very probably wrong.
- *Second law:* The only way of discovering the limits of the possible is to venture a little way past them into the impossible.
- *Third law:* Any sufficiently advanced technology is indistinguishable from magic.

If one thinks about the course of technological development over the past 200 years these tongue-in-cheek ways at looking at the world seem insightful indeed. Some of the world's most noted scientists and engineers have not been able to anticipate the extent to which we have been able to develop new technologies and do this at ever faster rates. Global innovation fueled by the Internet and global connectivity has brought us ever faster rates of development. The changes just between 1900 and 2000 in terms of cell phones, satellites, spaceships, genetic engineering, and miracle drugs have been amazing. Who is to say that in another century we will not have viable space colonies and smart cyborgs carrying out tasks from farming and manufacturing to even exploring the universe for us? Clarke has famously said that anything humans can imagine, there is a likelihood they can 1 day accomplish.

References

1. The First Private Space Habitat Is Here: The Bigelow Expandable Activity Module. http://bigelowaerospace.com/beam/. Last accessed Jan. 7, 2016.
2. Douglas Messier, "Private Space Stations Could Be a Reality by 2025," Space.com, August 25, 2015. http://www.space.com/30359-private-space-stations-reality-2025.html?li_source=LI&li_medium=most-popular. Last accessed Jan. 6, 2016.
3. The Bigelow Expandable Activity Module http://bigelowaerospace.com/beam/. Last accessed Jan. 7, 2016.
4. The Bigelow Proposed B330 Space Habitat. http://bigelowaerospace.com/b330/. Last access Jan. 7, 2016.
5. Joseph N. Pelton, *Orbital Debris and Other Space Hazards,* (2014). Springer Press, New York.
6. Interorbital Systems: http://www.interorbital.com/. Last accessed Jan. 10, 2016.
7. Daniel White, "Congress Wants NASA to Build a Deep Space Habitat" Dec. 30, 2015. http://www.spacesafetymagazine.com/press-clips/press-clips-week-52-53-2015/. Last accessed Jan. 10, 2016.
8. Doug Messier, "PISCES Embarks on Lunar Concrete Development Project" May 31, 2014. http://www.parabolicarc.com/2014/05/31/pisces/#sthash.sbJL9Onq.dpuf. Last accessed Jan. 10, 2016.
9. 3-D Printer for House Construction http://www.parabolicarc.com/2014/05/31/pisces/#sthash.u7G1Kw9Z.dpuf.

9

Governing the New Space Economy

Many believe that the international space treaties adopted a half century ago are yesterday's news. Today's most dynamic of the space entrepreneurs, such as Elon Musk, Paul Allen, Richard Branson, Larry Paige, and Peter Diamandis, look at a technology or a regulation that is five decades old and ask why we should be bound to follow rules from the "stone age." Those movers and shakers that are revolutionizing the New Space industry and embracing the new disruptive space technologies are focused on the future and not the past.

The problem is that countries of the world and especially spacefaring nations are bound by the international agreements that define "space governance" regardless of whether these rules and regulations were enacted 50 min or 50 years ago. Clearly in many ways the Outer Space Treaty and the four subsidiary international agreements have now become dated. Space technology has come a long way, and there are an increasing number of spacefaring nations and many new space services and industries that have appeared since these five agreements were negotiated back in the 1960s and 1970s. These documents are often thought to be ineffective or incomplete tools for growing and managing the twenty-first century New Space economy. This sense of failure in space governance comes from a variety of reasons that include the following:

Public Discontent

There is a general sense of discontent with all attempts at global governance of any type—whether local, national, regional or global. This includes a widely shared skepticism and even a sense of hopelessness that any of the U. N. initiatives in any sector can or will succeed. This is also seen in criticisms of the

European Union, NATO, and low approval ratings of national legislatures. Across the world there seems to be a lack of confidence in global governance mechanisms. This discontent and lack of faith seems to be out there regardless of whether the goal is keeping world peace, coping with refugee problems, or solving problems such as poverty or climate change.

Just Let Industry Do It

There is sometimes a parallel perception that private entities, some of which are international enterprises, seem to be succeeding where public space agencies have failed. Thus there is a growing view that we should just let industry "conquer" space and let governments simply stay out of it and stop wasting taxpayers' money. This corporate approach is seen as a replacement for international regulation, international agencies or even national space agencies. Just let the marketplace decide. Certainly the successes of the space billionaires have recalibrated public opinion about both space agencies and the effectiveness (or ineffectiveness) of international regulations related to space initiatives.

The Increasing Complexity of the Community of Space Actors

Further, the last 50 years have changed just about everything involving space. The political dynamics and scope of actors have gotten ever so much harder to deal with. The net result is that the attainment of any global consensus is harder than ever to secure. There are now many more countries on the world stage that did not even exist when the space treaties were negotiated. The number of countries with space-related programs has greatly increased from a handful to dozens. And perhaps most important the size, economic and strategic importance and range of space activities has greatly increased—along with the membership of the U. N. Committee on the Peaceful Uses of Outer Space.

New Issues to Contend with

There are a number of current issues that these space treaties and agreements did not anticipate and address. These issues include space traffic management and control, orbital space debris, a growing interest in the use of the protozone, or near space, and growing plans for a significant number of flights by suborbital trajectories for space tourism and, in time, hypersonic transport. These issues give rise to a variety of concerns that range from national security to public safety, environmental and health concerns, the integrity of vital public infrastructure in space, and concerns that problems like space debris might deny future viable access to space for all uses.

Legal Interpretations of the Space Treaties

Analyses have not strengthened these documents but perhaps made them more ambiguous in meaning. Various interpretations of these five treaties/conventions/agreements over time have accommodated various divergent viewpoints by different countries. The agreements today seem to actually pack less clout. The fact that new issues have arisen and various countries have argued to make the agreements less restrictive have at times served to weaken their provisions. Concepts and terms such as "global commons," "spacefaring nations," "celestial bodies," "space debris" and "space weapons" have, if anything, seemingly become more vague in their definition.

Today many feel there is a need for new definitions and perhaps new control or regulatory mechanisms related to outer space and especially with regard to the protozone because of all the new ways being found to use these regions just above commercial air space and just below orbital space. There have been efforts to create new space treaties and agreements or even codes of conduct, but always in the end these efforts have concluded without definitive results. Almost everyone seems to find fault with the Liability Convention, but no one seems to be able to agree on how to revise it. This process has led to the view that "model national laws" and administrative procedures and so-called "soft law" based on transparency and confidence building efforts may turn out to be the only way to move forward. This is because virtually nothing seems to actually happen in terms of globally agreed to new space treaties.

There have been moves calling for some form of space traffic management and control to contend with issues such as space tourism. Others have sought to contend with increasing levels of commercial launches to space—particularly with regard to large-scale constellations in low Earth orbit. Some are concerned with health and environmental concerns linked to outer space. Yet others have tried to focus on public safety concerns that arise from many different perspectives as diverse as orbital debris to the deployment and operation of private space habitats.

The bottom line is that no major new international consensus has been reached in the form of a space treaty or widely agreed international agreement for half a century. This inability to reach new agreements about how to contend with new problems in space is a major concern. There is a legitimate and even growing fear that the "outer space" frontier that is now being driven forward by new commercial space initiatives could become a sort of "Wild West," with no recognized regulatory authority for all the entrepreneurial space activities now on the horizon. Many new space ventures are reluctant to see space regulations enacted since they feel that such actions will be unduly restrictive. Yet ultimately, clear standards and efforts to create a level high frontier playing field may be just what the New Space ventures truly need.

The U. N. Committee on the Peaceful Uses of Outer Space that is responsible for international space policy and regulation has now grown to include over 80 countries. Yet most of the COPUOS member states are incapable of launching into space. What they do have in common is enormous diversity of political opinion, scientific knowledge, and contending views as to what is right and wrong with regard to space policy. A highly vocal COPUOS committee membership includes Canada, China, Cuba, most of Europe, India, Iran, Russia, the United States and Venezuela.

In light of this diverse membership it is easy to see why COPUOS is not always prone to consensus. The end result is often stalemate and putting off decisions to the next meeting. It seems that even efforts to agree on guidelines for the Working Group on the Long Term Sustainability of Outer Space Activities were postponed for a year at the COPUOS meeting in 2010. Sustainability of space is an item of some urgency. Yet a quibble over a few words in the draft terms of reference led to yet another year's delay at the very outset of this effort. Currently interventions from Russia and other countries at COPUOS sessions held in the spring of 2016 is now leading to further delays in efforts to find consensus on what to do about the long-term sustainability of space.

In short, in a total of 6 years, only baby steps have been made on voluntary guidelines to make space more sustainable. The bottom line is that it will apparently be ever more difficult for COPUOS to come up with consensus answers on space debris, space debris removal processes, strategies to cope with severe solar storms, changes to Earth's magnetic field, space traffic control and management, or virtually any other major pending space governance issue one might pick. An enforced rule of consensus agreement among the current 83 members of COPUOS with its politically diverse membership is an almost impossible feat. David Kendall will be president of COPUOS for 2016–2017. His experience as a top executive of the Canadian Space Agency and years as the Canadian representative COPUOS will serve him well, but getting absolute consensus at COPUOS meetings is a trial for the most judicious and tactful leaders.

In recent years, one of the few effective mechanisms that have helped to create new forms of space regulation has come from the coordinated efforts of space agencies. Space agencies have a clear and logical concern about the mounting threat of space debris and other space hazards such as coronal mass ejections. The Inter-Agency Space Debris Coordination Committee (IADC) actually took the lead in developing guidelines to restrict the new creation of space debris. These guidelines once agreed within the IADC were subsequently considered within the U. N. COPUOS, and with only minor changes accepted as "voluntary guidelines" there as well. In this case COPUOS took 18 years to polish off the voluntary guidelines on space debris. Lightning fast is clearly not in the vocabulary of the COPUOS forum (Fig. 9.1).

Clearly we need new ways to govern a global space economy. Without new rules and regulations we will not be able to survive and nurture the new needs of a twenty-second century world that depends on space as much as it does on Planet Earth. The remainder of this chapter thus explores new modes for achieving distributive justice and global corporate responsibility and maintaining a conflict free environment in space.

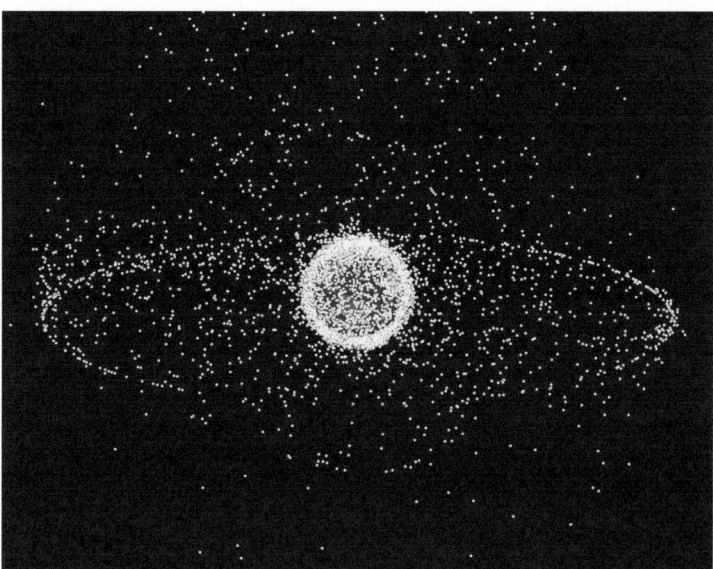

Fig. 9.1 Space Debris is one of the problems addressed by both the IADC and the UN COPUOS but the Problem Continues to Increase (Image courtesy of NASA)

Model Laws

Currently the lack of agreement on new space treaties and conventions has tended to shift new regulatory actions back to the national level. Today, at least France and the United States have created processes that serve to make the review process with regard to space debris mitigation "mandatory" rather than "voluntary" under its national legal or administrative processes and it is hoped that other countries will do the same. Model laws at the national level that are widely adopted can certainly be a part of the answer.

In cases where national laws on space regulation seem to run counter to international law, however, this can also serve to be a problem. Currently there is international concern about a new U. S. law that seeks to establish procedures and processes with regard to the future activity of space mining.

On November 10, 2015, Congress enacted the U. S. Commercial Space Launch Competitiveness Act of 2015 [1]. This bill, known as House of Representatives bill HR2262, passed both houses of Congress on a bi-partisan basis. Key elements of this key bill are outlined in Table 9.1 in this chapter. Significant innovations in this legislation under Title IV are the provisions that would seem to allow U. S. corporations or individuals to engage in this activity. Title IV includes the following language: "to provide U. S. citizens with the right to obtain resources from asteroids consistent with current laws and international obligations." This law also provides direction to the Executive Branch of the government to explore new ways to provide oversight of "commercial non-governmental activities in space."

These provisions in this new act appear to provide a legal basis under U. S. law for space mining of asteroids and other celestial bodies. It also seems to set the stage for expanded private enterprise activities in space that goes well beyond the commercial satellite applications undertaken today to provide communications, remote sensing, and navigation services. The prime question of the day is whether or not a number of other spacefaring nations will now proceed to enact essentially parallel legislation along the same vein.

At this time the initial reaction has been for several countries to express objections to the U. S. legislation. The Russian delegation to the U. N. COPUOS sessions in Vienna expressed the view that the new U. S. law should be considered contrary to the Outer Space Treaty, the Moon Treaty, and perhaps other international agreements. In the discussion views were expressed on both sides of the issue, and the final agreement was to place this item on the COPUOS agenda for the 2017 meeting [2].

Certainly the new U. S. approach seems to be consistent with precedents taken from the U. N.'s Law of the Sea, but does seem at odds with some provisions of

Table 9.1 Summary of the U. S. Commercial Space Launch Competitiveness Act of 2015

This act contains some of the following major provisions.
- Formally extends operation of the International Space Station from 2020 through 2024. President Obama announced in 2015 that he was extending it until then, but this will make it law. Canada and Russia have agreed with the extension; Japan and Europe have not publicly endorsed the extension yet.
- Extends the "learning period" for commercial human spaceflight through September 30, 2023. Under current law, the prohibition for the FAA promulgating new regulations for the commercial human spaceflight business expires on March 31, 2016.
- Extends third party indemnification for launch service companies through September 30, 2025. Under current law, the authority for the FAA to indemnify commercial space launch companies from certain claim amounts from the uninvolved public in the event of a launch accident expires on December 31, 2016.
- Directs the White House Office of Science and Technology Policy (OSTP) to assess and recommend approaches for oversight of commercial non-governmental activities in space. The 1967 Outer Space Treaty requires governments to authorize and continually supervise the activities of their non-governmental entities. (In response to the provision OSTP in April 2016 has recommended that the Department of Transportation—or in effect the FAA Office of Space Commercialization—review and provide licensed authorization for off-world space activities such as space mining, operation of private space stations, or on-orbit servicing or refueling of satellites or vehicles.)
- Establishes a legal right to resources U. S. citizens obtain from asteroids consistent with current law and international obligations. It directs the president to facilitate and promote space resource exploration and recovery.
- Provides a use policy for NASA's Space Launch System (SLS). SLS may be used for missions to extend human presence beyond low Earth orbit (LEO), for other payloads that can benefit from its unique capabilities, for government or educational payloads consistent with NASA's mission to explore beyond LEO, and for "compelling circumstances" as determined by the NASA Administrator.

the Outer Space Treaty and the Moon Agreement. The prime issue seems to be whether mining operations can be undertaken without a claim to "sovereignty" over the celestial object where such mining operations are to take place.

The issue of space mining and who might be able to do it and under what auspices is just one of the many issues involving space governance that are now in the vernacular "hanging fire."

Some of the top issues now pending include following, although this is far from a comprehensive list. These examples are illustrative of the fact that these are far from trivial issues. It is unfortunately the case that only when a disaster strikes and there are substantial losses of human life and huge financial interests lost that policy making bodies are finally persuaded to act.

SPACE TRAFFIC MANAGEMENT AND CONTROL. Perhaps the biggest item needing resolution is this. Why so? Because a lot of lives could be at stake. Some have suggested that the International Civil Aviation Organization (ICA0) headquartered in Montreal, Canada, should be responsible for coordinating such space traffic management and control. This would apply not only to Earth orbits all the way out to geosynchronous orbit but also the regions all the way down to commercial air space. In short there would be the equivalent of traffic cops for the protozone all the way through the various Earth orbits up to geosynchronous orbit.

Safety advocates argue that all that needs to be done is for ICAO to assume this role and carry out this function under its already existing charter. They suggest that ICAO could use its already established practice of developing "Standards and Recommended Practices" (SARPS) that would cover safety for the areas above commercial air space from 21 km all the way out to 35,870 km. Traffic controllers at the national level would do the bulk of the work, and ICAO would coordinate international patterns and standards of air and space safety practices. In this system commercial air space, the protozone (i.e., 21–160 km) and operations in Earth orbit as well would be coordinated by a single global entity, but national air traffic controllers would do most of the actual traffic controlling.

Others feel that the Chicago Convention of 1944 under which ICAO operates needs to be explicitly amended for it to be able to take on such a role. Yet others note that the issue is a bit more complicated. They point out that there are many other functions to be coordinated and potentially resolved before this would or should take place. Certainly the International Telecommunication Union (ITU) already has responsibility for allocation of radio frequencies. Then there are other areas of concern such as environmental pollution that would presumably involve the U. N. Environmental Program (UNEP) and the World Meteorological Organization (WMO). And what about regulations and standards for radiation danger that might involve the World Health Organization (WHO)?

Further there are potential issues of uses of the protozone and Earth orbit to deploy military or defense-related systems, and this would presumably involve the U. N. Office of Disarmament Affairs.

What this discussion points out is that when you begin to discuss space and space regulation, everything is up for grabs because there is not International Space Organization in charge of outer space. The result is that space-related matters are divvied up among scads of specialized U. N. agencies. When it comes to space there are dozens of parts and no whole. On top of this, what about the military units around the world that carry out space situational awareness activities to keep track of orbital debris elements that threaten key space assets and to be on the lookout for a missile attack? Clearly the issues of space traffic management and control and space situational awareness activities are closely linked, and even if the ICAO took over parts of this function military space situational awareness operations would continue.

THE MILITARY USES OF THE PROTOZONE AND EARTH ORBIT. Perhaps the only way to get regulators to focus on the need for safety regulation of the protozone is by making it a hot button issue directly linked to strategic military importance. Recently it has been recognized that there are strategic uses of both outer space and the protozone or, more generically, the stratosphere, all the way up to the area where satellites can stay in Earth orbit (20–160 km altitude). There is—quite simply—considerable and now growing concern about the various military or defense-related uses of both the protozone and outer space.

The world space law community recognizes that there is no one agency that is in a position to police these areas. In fact, it is even difficult to say definitely what is or isn't a space weapon. Who can say what can and cannot be done in these areas that is consistent with or in violation of the Outer Space Treaty? The problem is that the OST prohibits the deployment of space weapons, but this concept is not explicitly defined. Today there are satellites that are deployed that provide military communications—including tactical communications—defense-related surveillance, and space-based navigation and targeting. Further there are nuclear-armed missiles that follow suborbital trajectories and spacecraft that have nuclear power plants. The near-term future may see high altitude platform systems that could be deployed in the stratosphere or the protozone that may very well be used for military communications, surveillance or targeting. Such a protozone platform might even be utilized for the delivery of some sort of military strike or the positioning of a bomb or deploying a laser or directed-energy weapon system. Clearly a definition of both outer space weapons as well as protozone weapons are now lacking and constitute a major gap in space law. This gap is dangerous and needs to be addressed with urgency. This issue is, of course, also tied to the issue of space traffic control and management as discussed above. It seems impossible to resolve one of these issues without also addressing the other.

MITIGATION AND REMOVAL OF SPACE DEBRIS FROM EARTH ORBIT. If there is an overall theme that might be said to exist in the field of space and its regulation, it is that everyone has been quite careless in their handling of international space regulations and concerns for space safety. Satellites have been launched willy-nilly. Radio frequencies have been allocated without careful thought to future demand and longer-term needs. Debris has grown from modest to moderate to now almost out of control. Space applications have grown from almost a scientific curiosity to vital infrastructure on which millions if not billions now depend. This has all occurred without serious consideration of safety, future needs, interference, and clashes that may occur between commercial, civilian and military uses of outer space and the overseeing of space systems that may be vital to the long-term sustainability of human civilization. Some believe that we are on the very cusp of runaway and ever-cascading orbital debris. This condition could deny future generations cost effective access to space. Dr. Donald Kessler of NASA predicted in the 1970s the possibility of such a major calamitous event happening in the future, and now we need deliberate action to avoid such a runaway disaster [3].

ALLOCATION OF RADIO FREQUENCIES. The International Telecommunication Union has sought to bring order and equitable allocations to the world community in its processes and through its World Radio Conference. But the unparalleled growth in wireless and satellite communications systems has made a truly structured and rational approach increasingly impossible. The demand for ever more wideband services has moved satellite communications up from C-band, to Ku-band, to Ka-band, and now efforts are underway to test Q and V-band systems and perhaps ultimately W band and even terahertz communications systems. The meeting of ever mounting demand for broadband throughput will be an ongoing challenge. The efficient and cost-effective interconnection of various types of satellites is also an ongoing challenge, as well as protozone-band platforms and ground-based networks. Allowing growth and yet preventing interference between and among various types of sensitive services such as scientific radio telescope activities will become ever more difficult. The challenge is not only to expand frequencies to meet demand but also to increase radio frequency use with improved reliability and reasonable cost efficiency.

SOLAR POWER SATELLITE TRANSMISSIONS TO EARTH. The idea of creating solar power satellite systems that could beam back to Earth clean and green energy has been around for half a century. This idea was first proposed by Peter Glasser of Arthur C. Little in the 1960s. Today, finally, it seems to have evolved from an idea to a point where it might well become a reality. Enabling technology related to solar power conversion, power transmission systems, and more cost-efficient launch systems have all matured to the extent that these services might prove cost-effective in coming years.

However, the deployment of such systems involve more than having the right technology. Technical and safety standards and regulations will also be keys. The frequencies for transmission of the power will need to be decided as well as limits on the power transmission levels. The ground receivers must be spread out over relatively large areas so that the power density does not pose a health issue. The design of the ground receivers will be key, so that power is not reflected into space and unacceptable interference to communications and other satellites does not occur. There is also concern that the power transmissions from space do not interfere with communications satellites. If these issues can be overcome and power satellite systems can provide power in a cost-effective way, this could become one of the more significant of the space industries in terms of total revenues in coming years.

THE MINING OF OUTER SPACE AND USE OF SPACE RESOURCES. One of the currently hot topics is that which was once seen as a totally Buck Rogers concept of mining asteroids and the Moon. The formation of four companies to pursue space mining suggests that at least some people think that such enterprises can be viable in the not too distant future. Two companies, namely Deep Space Industries (DSI) and Planetary Resources Inc. (PRI) are developing deep space explorers to find asteroids to mine. Two other companies, namely Moon Express and Shackleton Energy, are focused on mining the Moon. Here the issues are many. Pending issues include safety issues related to remote mining operations and viable ways to return mined materials to Earth. There are questions as to whether the terms of the Outer Space Treaty and the Moon Agreement allow such mining operations to be undertaken as commercial operations, and whether celestial bodies and objects are off limits for private enterprise or rather these activities are restricted to some sort of international effort on behalf of the common heritage of humankind.

This is now an issue that is subject to some debate. Title IV of the U. S. Commercial Space Launch Competitiveness Act of 2015 seemed to say that extraction of materials from the Moon, asteroids or other celestial bodies can be undertaken without asserting sovereignty or ownership of the body itself. This U. S. interpretation of international law has been challenged by countries such as Russia that contend that one cannot mine bodies over which they do not command sovereignty and thus all celestial bodies are off limits [4].

These are just some of the many pending new regulatory and legal issues related to expanding or New Space industries. There are dozens of others where either safety standards are needed or national regulatory oversight is needed to meet requirements that flow from the Outer Space Treaty (OST) or the other international agreements that flow from it. Many of these activities, such as on-orbit servicing, active debris removal, commercial spaceplane operations, and space station deployment and operation will operate under U. S. regulatory actions. This is because many of these New Space activities are originating in the United States, and precedents are being set under

U. S. legislation or come from actions undertaken by the FAA Office of Space Commercialization (FAA-AST).

For those that are seeking insight into what changes are occurring in terms of new standards and regulatory guidelines, the best comprehensive source may well be the upcoming book *Global Governance of Outer Space*, which is a study that was conducted under the auspices of the McGill University Institute of Air and Space Law (see below). This study provides useful updates on changes to space standards and regulations as well as recommended actions that might well help to illuminate the way forward [5].

Comprehensive Findings of the Study on Global Governance of Outer Space

The Global Governance of Outer Space conference sought to examine the full extent of current and planned activities in space. These various space activities now exceed a quarter of a trillion (U. S.) dollars per annum and impact the global economy to a much larger extent than before. Global navigation satellite services and satellite communications impact almost every economic aspect of the global economy to some degree or another. Table 9.2 seeks to summarize the broad findings of this landmark study and where new standards,

Table 9.2 Key areas of space activities and areas where new standards or regulatory actions are recommended for consideration and action

Area of Concern	Areas of Possible Future Action on Standards or Regulation
1. The Outer Space Treaty and its limitations	The treaty does not define a number of key terms such as celestial bodies, space weapons, etc. It does not recognize the possibility of truly independent and private space actors and puts undue responsibility on nation states and "launching state."
	The major pending issue appears to be with regard to conflicting ideas about "sovereignty" over off-world assets versus the act of resource extraction, a more precise definition of what is a "celestial body" and precise interpretation of the OST.
2. The Liability Agreement and its limitations	The Liability Agreement does not anticipate the problem of significant orbital space debris and does not create any incentive for space debris removal. It also did not anticipate the extent to which future space activities may well involve many more private enterprises in space and relating this activity to the responsibility of the "launching state."

(continued)

Table 9.2 (continued)

Area of Concern	Areas of Possible Future Action on Standards or Regulation
3. Satellite telecommunications	There is a need for allocation of additional radio frequencies; reduction in orbital congestion; coping with orbital debris and related concerns with new mega-low Earth orbit systems and their safe operations; reduction of jamming of satellite communications; and improved frequency coordination with high altitude platform systems (HAPS).
4. Satellite broadcasting	There is a need for allocation of additional radio frequencies; coordination with other satellite systems to reduce interference; improved tariffing arrangements to allow access to international satellite broadcasting.; and improved utilization of S-band allocation for educational broadcasting.
5. Remote sensing, Earth observation and meteorological satellites	There is a need for improved provisions for real-time sharing of remote sensing data to respond to disasters; improved standardization of formatting of remote sensing data; and improved standards to allow real-time downloading of meteorological data.
6. Global navigational and timing satellite services	There is a need for greater resilience against interference and jamming of GNSS transmissions; backup to loss of timing from GNSS for critical services such as banking transactions, synchronization of the Internet and aircraft safety operations; and compatibility standards among GNSS satellites.
7. On-orbit servicing and active debris removal	There is a need for nation-state licensing of on-orbit servicing operations; liability insurance provisions for catastrophic space losses; protections against accidents involving active orbital debris removal; safety standards against accidents involving refueling, in-orbit repairs, etc.; and new guidelines, standards and regulations concerning the reuse and recycling of defunct satellites and whether "salvaging" of satellites that have become space debris is legal or can be agreed among concerned parties and especially what are the associated liability arrangements.
8. Private space station deployment and operation	Who can authorize and license the operation of private space stations or habitats? Are national licenses as well as international regulations are required? Are different safety standards, licensing and other arrangements necessary for unmanned and manned space facilities?
9. Deployment of large scale satellite constellations	Who beyond national government licensing authorities and the ITU—with regard to frequency allocations and orbital positioning—are required to authorize so-called mega-low Earth orbit constellations? What are the special requirements for the operation and de-orbit of satellites from large-scale constellations? At what size and complexity of operation should a constellation be subject to special controls and regulations? Should the ITU, COPUOS or another entity such as ICAO—in the context of space traffic management and control—be involved in authorization, oversight, or other forms of control?

(continued)

Table 9.2 (continued)

Area of Concern	Areas of Possible Future Action on Standards or Regulation
10. Space traffic management and control	Should there be some form of space traffic management and control? Who should manage the interface between the control and safety of air space and space-related operations? Who should manage the protozone? Does the region between 20 km and 160 km altitude have sub-zones of operation and control with regard to free passage and how is that managed? Do the precedents from the Law of the Sea with regard to 12 nautical miles, 24 nautical miles and 100 nautical miles have a useful possible effect as a precedent with regard to the regulatory control of the protozone both with regard to areas over nation states and over international waters and polar regions?
11. Issues related to the use of the protozone	Does the same entity that is responsible for space traffic management and control play a comparable or parallel role with regard to the protozone? Should high altitude platform systems, robotic freighter planes, spaceplanes, hypersonic transports and stratospheric balloons be under the same control authority and will different rules apply to each type of object that might enter, fly through, or remain essentially in a single location?
12. Global space security and military uses of outer space	There are continually evolving military or strategic applications of outer space, Earth orbital space, and the protozone, and these might involve surveillance, remote sensing and Earth observation, tactical and non-tactical communications, weather monitoring, traffic controls and targeting, and on-orbit positioning and servicing, satellite positioning and de-orbit. The lack of an internationally agreed process to control these operations and coordinate the actions of various military and defense entities suggests a great potential for unintended misunderstanding and triggering of military actions due to misperception of intent. Areas of uncertainty and possible miscommunications are dangerous. Space situational awareness is becoming more and more difficult, costly and dangerous as the level of activities in space and the protozone increase. This thus becomes one of the most urgent for agreed modes of coordination and negotiation of international regulatory actions.
13. Space environmental and health issues	There are legitimate concerns about increased use of outer space, higher levels of stratospheric pollutants (which have perhaps a 100 times greater negative impact than at sea level), concerns about certain particularly "dirty" solid fuel particulates, plus other issues such as sonic booms involved with spaceplanes and hypersonic transport, and radiation levels for passengers in these new high altitude flight patterns. In addition these concerns it is not clear as to whether they are to be addressed at the national regulatory level, by the U. N. Environmental Program, the World Meteorological Organization, the World Health Organization, or some other entity such as the U. N. COPUOS and its working group on the Long Terms Sustainability of Outer Space Activities.

(*continued*)

Table 9.2 (continued)

Area of Concern	Areas of Possible Future Action on Standards or Regulation
14. Asteroids, comets, solar storms and other space hazards	The U. N. General Assembly and COPUOS need to expand their efforts that have begun through the International Asteroid Warning Network (IAWN), the Space Mission Planning Advisory Group (SMPAG) and the COPUOS Working Group on Long Term Sustainability of Outer Space Activities. There needs to be an InterAgency Advisory Committee on Cosmic Hazards and Planetary Defense that includes participation by groups such as the B612 Foundation and the Safeguard Foundation. Space agencies should adopt strategic plans with specifically targeted capabilities to identify cosmic threats and planetary defense capabilities.
15. Long-term sustainability of space	Closely tied to efforts to identify cosmic hazards and planetary defense are related efforts with regard to combatting and reducing orbital space debris (with active debris removal) as well as efforts related to addressing climate change, environmental pollution from space-related activities, solar radiation and the ozone layer, and concerns related to shifts in Earth's magnetic poles and reduction of the Van Allen Belts' ability to protect again coronal mass ejections. This should be addressed by U. N. COPUOS and the InterAgency Advisory Committee established to address cosmic hazards and planetary defense.
16. Space launch services	Currently space launch services are still largely regulated at the national level with regard to safety, debris mitigation, frequency assignments and launch registration with the United Nations. This is likely to continue since launch operations are closely tied to national defense. Concerns related to planetary defense, climate change, orbit space debris, orbital congestion, frequency coordination and jamming will require improved international cooperation. This might be addressed by an InterAgency Coordinating Committee. For private space launches, the Commercial Spaceflight Federation or the International Association for the Advancement of Space Safety (IAASS) and Secure World Foundation (SWF) should provide support in developing international guidelines and coordination.
17. Human space missions	In the past all human space missions have been conducted by national space agencies (i.e., NASA, Roscosmos and the Chinese National Space Agency). The advent of private spaceplane flights, private space habitats and other private space initiatives requires new oversight, safety standards, and regulation of both spaceport safety and related space traffic management and control. This could be assisted by a global space safety institute to help develop standards. A global entity such as ICAO should assume responsibility for coordination of safety standards for private and public space traffic management and control. Health standards for radiation also need to be globally agreed. Globally accessible space situational awareness information needs to be available to be alert to orbital debris dangers and possible conjunctions with space objects at all times. All of the same concerns apply to human space activities in the protozone.

(continued)

Table 9.2 (continued)

Area of Concern	Areas of Possible Future Action on Standards or Regulation
18. New Space manufacturing and processing	The advent of space mining and resource extraction will quite likely be accompanied by efforts to engage in processing and even manufacturing of launcher fuels and products in space. This will at a minimum require the adoption of safety standards and some form of licensing oversight or regulatory process required under the OST. At least initially these efforts will likely involve oversight by nation-states (including relevant private activities carried out under "launching countries"). New liability and other provisions will also likely need to evolve over time. Organizations such as the IAASS, the IISL, COSPAR, and the International Academy of Astronautics may be able to assist in this regard as well as in other New Space initiatives and issues addressed earlier.
19. Space Colonization and Migration	In light of national space agency and private initiatives to go to the Moon and Mars and to eventually establish off-world outposts, the time has come to consider rules of safety, resource extraction, outpost protection, mutual aid, traffic control, pollution, even juridical process for such off world outposts that could eventually lead to colonization and even migration. At this time COPUOS and launching states will take the lead on these matters, but the other organizations listed above.
20. Sharing the bounties of the cosmos and common heritage of humankind	The Moon Agreement specifies that celestial bodies are not subject to being claimed as the sovereign territory of any nation and are like the oceans and Antarctic part of the "Global Commons" and thus essentially the "Common Heritage of Humankind." Also evolving elements of international law are increasingly focusing on the rights of future generations. The crunch issue is what might be called the "balance" dilemma. This dilemma is how to balance the costs and new technological needs required for the future exploration and exploitation of outer space versus the designation of space as part of a "commons," equally available to all. Various suggestions of global entities patterned after Intelsat, Arianespace, or other international institutions have not jelled. Private ventures in the United States to pursue space mining and the new U. S. Commercial Space Launch Competitiveness Act of 2015, plus new U. S. processes to license such activities by FAA-AST seems to be setting up a clash of national and private interests versus the provisions of the OST and the Moon Agreement. These conflicting views may be able to be resolved some sort of balancing compromise. These issues are a complex blend of national security, political and legal concerns and as such will remain difficult to resolve for some time to come. Discussions in COPUOS will thus likely continue for some time to come until a political/quasi-legal agreement is negotiated among prime stakeholders.

(*continued*)

Table 9.2 (continued)

Area of Concern	Areas of Possible Future Action on Standards or Regulation
21. Need for key definitions related to Global Space Governance	Progress with regard to Global Space Governance will likely continue to be made point by point in the various specific areas presented above. But progress can be made if agreement could be reached either formally or even informally within COPUOS, InterAgency coordination processes, the U. N. Office of Disarmament Affairs, or the International Courts addressing space law. This would particularly be the case with regard to a consensus definition of such concepts as space debris, space resource extraction, space weaponry or perhaps with regard to clarification of the scope or size of a celestial body.
22. Modifications to the Outer Space Treaty or Liability Agreement	Another area of general progress would be some modest changes, amendments or clarifications to the OST and Liability Agreement that recognizes the broader role of private entities in space activities, to address the need and incentives for orbital debris removal, or other issues identified above.

Note: This chart was prepared by the author as a supplementary research effort in support of the Global Governance of Outer Space Investigation. It represents a separate set of findings and recommendations, since the findings precede the final report

regulations or laws might be needed assist with new space services or activities in coming decades.

Conclusions

The future of outer space activities is going to be determined in many ways by new policies, regulations, safety and health standards, model national laws, and transparency and confidence building measures. The difficulty of creating broad agreements through new space treaties or international agreements will remain extremely difficult for many reasons. These reasons include: (1) the rise of many actors participating in space-related activities; (2) the diversity of political views and concerns among the members of COPUOS as well as the expanded membership of this body; (3) the rise in the strategic importance, number, and value of space assets related to national defense; and (4) the possibility that New Space capabilities developed for space situational awareness, on-orbit servicing, active space debris removal, planetary defense, or space commercialization could also be used as space weapons.

Despite these difficulties, the informal agreements, private space initiatives, and new technology can help. The New Space technology is now resulting in low cost launchers and more efficient space operations. All of these innovations can still allow significant progress in many areas of improved space governance.

However, much more is now moving the cause of space governance forward. The McGill University multi-disciplinary study of space law and regulatory action known as the "Global Governance of Outer Space" is an unprecedented effort to bring technologists, space lawyers, practitioners and relevant space regulatory officials together to chart a new path forward. Future prospects will be aided by the UNISPACE + 50 Conference in 2018, the work of the U. N. COPUOS and especially its Working Group on the Long Term Sustainability of Outer Space Activities, the U. N. Office of Disarmament Affairs, ICAO, and InterAgency consultant groups such as the IADC. Further assistance can come from organizations such as the IAASS, the Secure World Foundation, the B612 Foundation, the Commercial Spaceflight Federation, COSPAR, the International Academy of Astronautics and many others. These international initiatives, often because they are informal and non-governmental yet international in scope, are all helping to move the cause of global space governance forward.

The conclusion is actually quite simple. The new gold rush will not happen unless we can unclog the wheels of global space governance so as to allow key innovations and progress in space. The power of New Space initiatives and entrepreneurial innovation is strong, but this power cannot be unlocked if there are no rules of the road to allow fair competition, equitable access, and safe, secure and viable access by all to the bounties of cosmos.

References

1. Senate Passes Compromise Commercial Space Bill – UPDATE, November 11, 2015. http://www.spacepolicyonline.com/news/senate-passes-compromise-commercial-space-bill.
2. Jeff Foust, "Mining Issues in Space Law" May 9, 2016. http://www.thespacereview.com/article/2981/1. Last accessed on May 21, 2016.
3. Donald Kessler, "We're Polluting Space Faster Than Nature Can Clean it Up," *Huffington Post*, Sept. 24, 2013. http://www.huffingtonpost.com/2013/09/24/space-junk-donald-kessler_n_3983899.html. Last accessed May 21, 2016.
4. Ram Jakhu, Joseph N. Pelton, and Yaw Nyangpong, *Space Mining and its Regulation* (2017). Springer, New York.
5. Ram Jakhu and Joseph N. Pelton, *Global Governance of Outer Space* (2017). Springer, New York.

10

Policing the Gold Rush in the Skies

Introduction

The most difficult question about whether the new gold rush will succeed or fail is not what most people might think. It is not whether we are smart enough to invent new space technology to make these ventures profitable. Here we will succeed. It is not whether we can work out agreements to let New Space commerce proceed to reap the vast riches that populate the Solar System. People doubted the wisdom of the Louisiana Purchase, the future market for computers, or even the need for a patent office because all the useful inventions had supposedly already been conceived. But people who bet against future progress tend to be wildly wrong.

No, the top question is whether we can create a process to "police and enforce" the fair, equitable and sustainable efforts associated with the New Space commerce. It will be incredibly difficult to track and police the movement of people, manufacturing, processing and living habitats beyond the grasp of Earth's gravity well. Here is a key challenge that humans will find somewhere between baffling and completely daunting. The *Star Wars* movies tell of space smugglers, galactic bandits and illegal trade across the far reaches of outer space. Someday such fiction could indeed become reality. What we do know is real is the difficulty of enforcing regulations across the reaches of cosmic space. In the real world of physics, the speed of light is a true limit. Warp speed is for science fiction authors only, not for engineers and rocket scientists.

Living, working and even surviving in the vast reaches of outer space pose great challenges. People in space will be dependent on machines, smart robots, cybernetic organisms, and life support systems that will remain crucial for survival for some decades to come. Only if humans are smart enough and environmentally clever enough to terraform Mars or build vast space colonies in the vacuum of the cosmos such as envisioned by Gerard O'Neill will humans be able to live and breathe in space free of life-sustaining space suits. This will not come for many generations yet to be. People will not want to live their lives millions of miles away from hospitals, health care units, police, fire, emergency medical responders, armed forces and civil defense resources, and courts of law. The challenges of living a productive and healthy life for space pioneers could resemble the days of remote settlers who were isolated from the benefits of a thriving social, economic, legal and civil society.

In such an isolated and remote existence, will we be able to avoid regressing back to the same petty squabbles we constantly find unavoidable here on the 6-sextillion-ton mud ball we call Planet Earth? Geopolitics is hard. Exploiting the resources of the Solar System in a fair, balanced, sustainable and non-confrontational manner will perhaps be much, much harder. To think that this can be done without policing and enforcement mechanisms with consequences is naïve. Some form of governance will be needed. This chapter discusses the hard realities of cosmic policing mechanisms and the legal and regulatory systems that may be required for space colonies and habitats in space to be possible and practical.

The Outer Space and Moon Treaties: Global Commons Applied Elsewhere

The Outer Space Treaty, formally the Treaty on Principles Governing the Activities of States in the Exploration and Use of Outer Space, including the Moon and Other Celestial Bodies, forms the basis of international space law [1]. This treaty that entered into force in October 1967 half a century ago set forth a number of high-minded and what some would say are idealistic goals. These goals are summarized by the U. N. Office of Outer Space Affairs as presented on their official website. These key goals from the treaty include the following points that might be characterized as ideals in which space exploration can ethically be carried out:

- The exploration and use of outer space shall be carried out for the benefit and in the interests of all countries and shall be the province of all mankind;
- Outer space shall be free for exploration and use by all states;
- Outer space is not subject to national appropriation by claim of sovereignty, by means of use or occupation, or by any other means;

- States shall not place nuclear weapons or other weapons of mass destruction in orbit or on celestial bodies or station them in outer space in any other manner;
- The Moon and other celestial bodies shall be used exclusively for peaceful purposes;
- Astronauts shall be regarded as the envoys of mankind;
- States shall be responsible for national space activities whether carried out by governmental or non-governmental entities;
- States shall be liable for damage caused by their space objects; and
- States shall avoid harmful contamination of space and celestial bodies [2].

In the 1970s the Outer Space Treaty was supplemented by what is briefly referred to as the Moon Agreement. This supplement to the Outer Space Treaty is formally known as Agreement Governing the Activities of States on the Moon and Other Celestial Bodies. This international agreement was adopted by the General Assembly in 1979 in Resolution 34/68. It was not until June 1984, some 5 years later, however, that the fifth country, Austria, ratified the agreement. This signature finally allowed it to "enter into force" in July 1984. While the Outer Space Treaty has been widely ratified, the Moon Agreement still has been formally ratified by only about 15 countries as of 2016. Key language from the Moon Agreement, that none of the major spacefaring nations have yet ratified, is as follows:

> *The Agreement reaffirms and elaborates on many of the provisions of the Outer Space Treaty as applied to the Moon and other celestial bodies, providing that those bodies should be used exclusively for peaceful purposes, that their environments should not be disrupted, that the United Nations should be informed of the location and purpose of any station established on those bodies. In addition, the Agreement provides that the Moon and its natural resources are the common heritage of mankind and that an international regime should be established to govern the exploitation of such resources when such exploitation is about to become feasible* [3].

Not only have none of the major spacefaring nations signed the Moon Agreement but there is rather wide disagreement about what the "common heritage" of mankind means, especially when it comes to the possible settlement and future "exploitation" of resources from space.

What is known is that over the past 40 years since the Moon Agreement was negotiated and agreed to the world has changed in significant ways. World population has grown from 5 billion to nearly 7.5 billion and seems headed to a level somewhere between 10 billion and 12 billion by 2100. No one knows if the global population will stabilize by that time or not. The world has begun to experience severe water shortages, and many metals and rare earth substances have grown more and more scarce. Some 25 million people have relocated from the Sahel region of Africa due to water shortages and drought. The mean temperature of the planet has increased some 2 °C and is continuing to rise along with a growing amount of greenhouse gases in the atmosphere. The reaction to these changes has included a new appreciation for recycling of resources and sustainability. There is also increased support for improved automobile fuel efficiencies and cleaner emissions from all types of vehicles, from cars and trucks to jet aircraft. A world that once viewed the future as an endless cornucopia of wealth, growth and prosperity has come to recognize that the warnings from Malthus, the Club of Rome, and James Lovelock—of "Gaia" fame—need to be taken seriously. In this changed world there are emerging new views that space resources, solar energy, along with the resources of the ocean may be needed to rescue humanity from its overconsumption of Earth's resources.

Part of this twenty-first century recognition of fundamental change has been the recent creation of four companies aspiring to engage in space mining and recovery of resources from space. The four U. S.-based companies are Deep Space Industries, Planetary Resources, Inc., Moon Express, and Shackleton Energy Company. All have different business plans and are pursuing their aspirations via new types of technology. Two of these companies are focused on the Moon and two are focused on asteroids. But all of these companies are in lockstep in terms of rejecting the aspirational ideals included in the Outer Space Treaty and the Moon Agreement. They have intensely lobbied, along with the Commercial Spaceflight Federation, for a new legislative agenda that permits new approaches to space commercialization.

The results of their legislative initiative is a new U. S. law known as the 'Space Resource Exploration and Utilization Act of 2015, which is Title IV of the Commercial Space Launch Competitiveness Act, which for brevity's sake can just be referred to as the "act."

Space Resource Exploration and Utilization Act of 2015—A New Space Commerce Paradigm or the Start of Space Conflicts?

The U. S. Congress passed this new act at the very end of its sessions in December 2014, and it was promptly signed into law. There are laws that are sweeping in scope. Such laws are considered to produce a "sea change" for the future. This act, by its broad new provisions, may prove to be not a "sea change" but an "outer space change" for the entire future of space exploration and off-world activities. At sessions attended by space lawyers around the world this is the constant subject of discussion—and argument. Although the reality of space mining is perhaps two decades away, this is what the world of space lawyers are focused on today. Here are what the provisions of this act specify:

> *The President, acting through appropriate Federal agencies, shall –.(1) facilitate commercial exploration for and commercial recovery of space resources by United States citizens; (2) discourage government barriers to the development in the United States of economically viable, safe, and stable industries for commercial exploration for and commercial recovery of space resources in manners consistent with the international obligations of the United States; and (3) promote the right of United States citizens to engage in commercial exploration for and commercial recovery of space resources free from harmful interference, in accordance with the international obligations of the United States and subject to authorization and continuing supervision by the Federal Government [4].*

This new U. S. law when signed into force, in a single instant, created a new direction for space exploration and New Space commercial initiatives. The provisions at the very end of this law stated that the provisions in this act were considered to be "in accordance with applicable law, including the international obligations of the United States" [5].

And further stated: "It is the sense of Congress that by the enactment of this Act, the United States does not thereby assert sovereignty or sovereign or exclusive rights or jurisdiction over, or the ownership of, any celestial body."

Despite these assertions there are many in the world space law community that interpret this new U. S. law to be assertion that the United States and New Space companies seeking to engage in space mining have a free pass to obtain resources from space and sell them on a commercial market free from the constraints of such considerations as sharing such exploited resources as a part of the "common heritage of mankind" [6].

Instead of a future based on "space colonies" constituted under a U. N. charter where there is a universal sharing based on a "global commons," where everyone is cooperating on the basis of the "common heritage of mankind," the new U. S. law can be seen to envision a much different space future. The image of the United Kingdom's past colonial enterprises such as the East India Company, which established overseas commercial empires, could be invoked as a model for the now emerging New Space paradigm.

And there is certainly a valid concern here that must be addressed. There does seem to be a coming space future where certain spacefaring companies and nations develop the technology, finance the mission, and perhaps risk the lives of astronauts and billions of dollars of equipment to launch vehicles and "smart machinery" to reclaim resources from space. These entrepreneurs and investors do feel that there are legitimate concerns about what is a fair system here. How and to what extent should these space miners be forced to share those resources mined in space with those with "no skin in the game."?

A "global commons" only seems to be legitimate and fair when there is a global investment of resources. Global sharing seems to involve a quid pro quo that currently seems missing from the equations devised in the Outer Space Treaty and the Moon Agreement that was negotiated in an age when reclaiming resources from space was merely a theoretical concept rather than a possibly dawning reality. In theory, sharing is easy. Sharing in reality involves a different set of rules.

This returns us to the issue of policing and legal enforcement in an off-world environment. As space security analyst Daniel Porras has suggested: "Outer space security and its enforcement is remote, filled with danger and vulnerability and ultimately—unruly. In short it is likely to be like the Old West in the United States."

We can certainly now envision a future where a New Space commercial venture creates an outpost on an asteroid millions of miles from Earth and extracts volatiles (such as water) and platinum and other rare metals from this remote space rock. The question arises with such a development as to who governs and keeps tabs on this far-distant operation? The even more challenging question is how would this be possible? The fundamental issue in question is, who is physically able to enforce laws and regulations in space, especially when it expands into an ever broader space arena?

The U.S. act acknowledges that it has obligations under the Outer Space Treaty and that includes oversight of a U. S. entity that would engage in space mining. Appendix 1 in this book provides the current recommendation from the White House as to how that oversight and licensing of efforts such as space mining would be carried out in the United States.

Earth orbits are already very challenging to manage. Management of Earth orbits requires keeping track of orbital locations—for satellites, debris and even missiles. It involves monitoring frequencies and interference, avoidance of space debris, and more. The U. S. government will likely invest over $8 billion in capital equipment to install the new S-Band Radar Fence that will range over 1000 km in length along the Micronesia atoll. This is just to track objects in low Earth orbit, and it doesn't include operating costs. The concept of yet another S-Band Space Fence in western Australia is under study.

There are ambitious plans by the B612 Foundation to build and launch an infrared telescope called the Sentinel that can survey the skies for potentially hazardous near Earth asteroids in the region of space from the orbit of Venus to Earth, but this is for detecting the presence of dangerous space rocks and certainly does not constitute a means to monitor remote operations. This capability is in no means anything like a monitoring device that could be used as a means of monitoring and "policing" remote human activities in space. A much more comprehensive, expensive and complex system would be needed for such a purpose (see Fig. 10.1).

It has been a challenge to get agreement on space traffic management and control for the region above commercial air space, which includes protospace and low, medium and geostationary orbits. What rules can be agreed and enforced beyond the realm of Earth orbit? How feasible would it be to extend the rules to include perhaps the unique and valuable Lagrangian points? What about tracking and control equipment that extend to the Moon and perhaps creating a traffic control system for orbits around the Moon? Once you start thinking about space traffic control and management it becomes difficult to define where you begin and how extensive of an area is to be monitored and controlled.

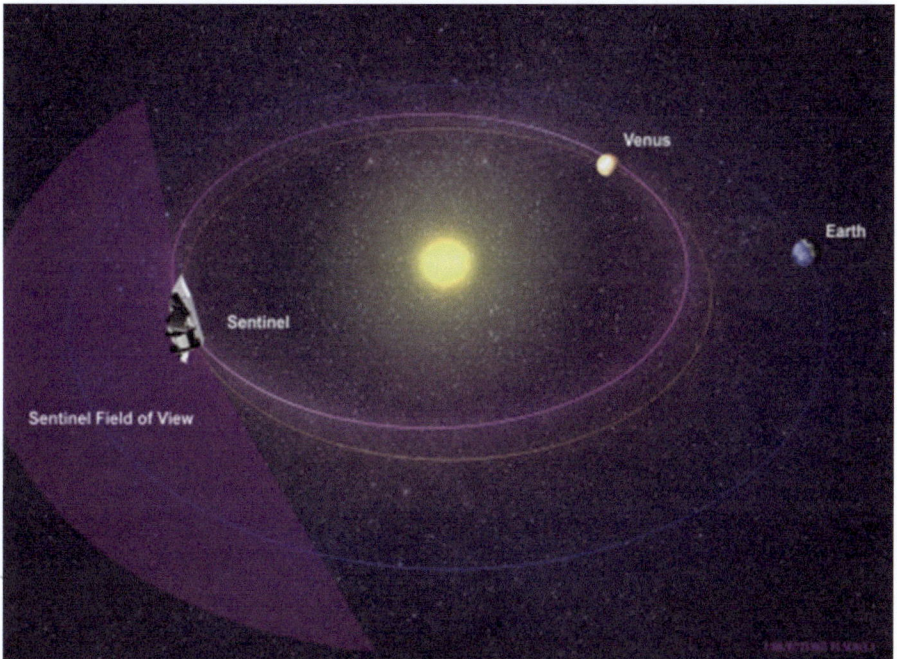

Fig. 10.1 The Sentinel infrared telescope that would seek to find potentially hazardous asteroids (Image courtesy of the B612 Foundation.)

The specific questions become, who and/or what enforcement agency will be able to maintain control of these regions and supervise the "lawful activities" that might be carried out there? Who would pay for such an oversight and "policing" operation? How would "violators" be brought to justice and penalties enforced? When these issues are no longer considered as theoretical issues, but brought down to specifics such as who pays for detecting violations and who is in charge of enforcement, the practical difficulties of regulating and policing outer space start to be seen as incredibly difficult.

History has repeatedly shown that laws and regulations only have effective jurisdiction if there is some form of incentive for observing and clear penalty for not observing the rules, i.e., an enforcement power. There are an estimated million near Earth objects that are some 35 m in size or larger that come within 0.05 astronomical units (4.5 million miles, or 7 million km) of our planet. The practical aspect of trying to create some form of policing or regulatory system for near Earth objects alone is a daunting task. How does one police millions of pieces of rocks whizzing around the Sun? (see Fig. 10.2).

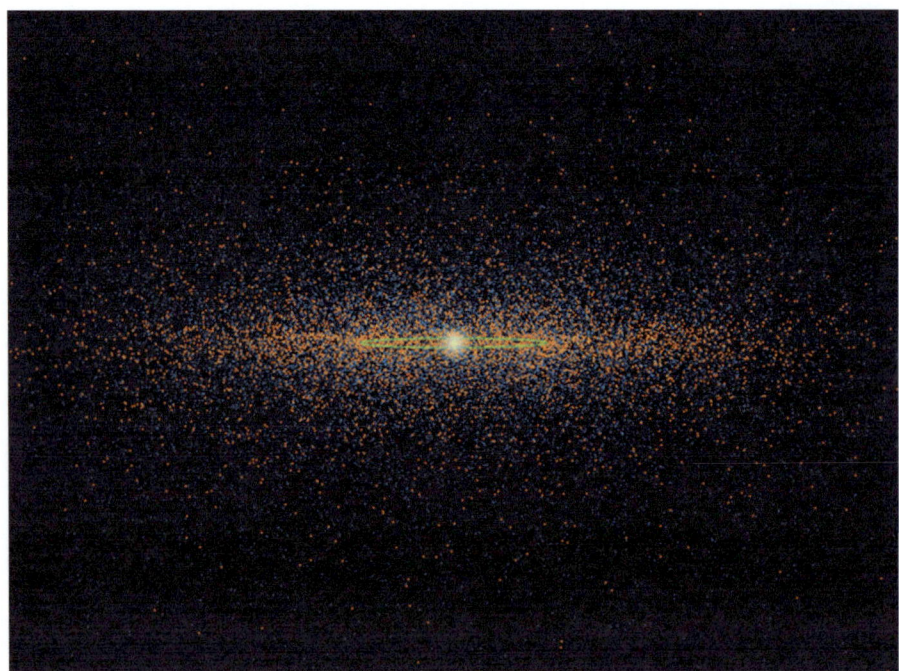

Fig. 10.2 How do we police space mining missions on potentially millions of near Earth asteroids? (Image courtesy of NASA.)

New Governance Concepts

Some believe that the exploration, settlement and utilization of space is a possible basis for a more hopeful future in which new forms of societal organization and self-control might evolve. Noted legal scholar and thinker George Robinson, working with a number of scholars assembled via the Smithsonian Institution, has conceived of a new type of organizational structure for future space activities based on the idea of fusing together governmental, commercial and academic systems into what he has called a "tripartite space governance unit."

Robinson, in describing his ideas about future types of space-based governance and control systems, starts with a very basic thesis that many might argue with today. He suggests that off-world activity is essentially a quest for survival. As he has put it: "The principal, if not sole, objective of all space activities is to facilitate and enhance incrementally the exploration, migration, dispersal, adjustment, and off-Earth settlement of the humankind genome to ensure the survival and evolution of its 'essence'" [7].

Indeed Robinson contends that this is the evolutionary essence of all biota that has evolved on Planet Earth and presumably any other biological form that might have evolved elsewhere in the universe. He contends that if there were to be new settlements elsewhere they should seek to combine the strengths of three modern pillars of today's society: (1) governmental systems (i.e., executive administration, legislation, policing and enforcement of law, and jurisprudence); commercial, industrial, and banking investment systems that include business innovation; and educational and research systems. These institutions do have strong roles in most democratic nations today, but this is not formally recognized in our governance system. This is what Robinson has called his tripartite governance system for off-world environments. He sums up his rationale as follow:

> If there is to be an effective and proper use of space and its resources, particularly as they relate to the enhancing and facilitating of human space migration through reliance on governmental, private commercial activities and contributions, and basic/applied research capabilities and interdisciplinary education and training, it must be in the form of a tripartite global entity. In the process of forming such an entity, there must be a clear understanding of precisely what the governmental role will be, and particularly in the context of allowing and even helping the private industrial and investor communities understand the opportunities for investment in space migration-related systems and services – particularly the need for developing a broad range of potential sources of private investors [8].

What is insightful about Robinson's ideas is that today it is actually private investors and private entrepreneurs—as much as anybody—that are leading the drive to undertake space mining, to create efforts that would go to the Moon, Mars or other off-world locations.

Of course we do not have to go to legal scholars or political scientists or even space agencies to consider what might be the best way to establish off-world colonies and how to best police and govern them. There are also literally thousands of science fiction stories that can suggest ways that off-world colonies or habitats or scientific settlements might be established, run, policed and economically sustained. Some are harmonious democracies that are able to sing "Cumbaya" together. Others become absolute dictatorships with iron rule. Some of the fictional colonies over time deteriorate and form into competing or warring tribes, such as more or less occurred with the unsuccessful "Biosphere 2" experiment. Others become a part of a vast galactic "empire," where police or soldiers are magically deployed around a galaxy without regard to the laws of physics. In these future cosmic worlds there are such unlikely capabilities as warp drives or interstellar transport via wormholes to span light years of distance in an instance.

Despite the great diversity of what might be, it turns out that the key questions to be faced are actually rather few. One question is, how are the economic investments needed to sustain off-world activities actually achieved? Or put differently, can off-Earth enterprises be developed that are self-sustaining and profitable? Another key question is, can off-world colonies and habitats be self-governing and policing, and if so, under what authority? Localized governance at a specific site in space? Governance from Earth? Or by some combination of processes such as the captain on a ship at sea who must report to superiors when he or she reaches port again?

Evolutionary Development Geared to the Needs of the Time

It seems most likely that such issues and concerns will follow a practical, common sense, evolutionary approach with increasing sophistication of approach evolving over time. In the case of private space habitats in Earth orbit or space missions of short durations there will be designated "captains" of such missions that will be charged with command and control, but subject, just as is the case with maritime law, to behavior review when the captain and crew and passengers return to Earth. Over time, as missions become more complex, with smart robotics, longer term operations on asteroids, the Moon, Mars, or artificial space colonies, then systems for governance, policing and enforcement will need to be developed to protect human rights and settle disputes or conflicting claims. Practical solutions to practical problems will be found. The time of global consensus that resulted in the negotiation and agreement on the Outer Space Treaty and other subsidiary conventions and agreements has apparently come and gone.

Today useful mechanisms for outer space agreements actual abound. There are key things such as model national laws, plus codes of conduct, transparency and confidence building measures, bilateral or multi-lateral negotiations and agreements. All these things can and do play a key role. There are also many other ways for the coordination and negotiation of space activities to be carried out.

The list of current mechanisms that might be used to coordinate international agreements in the area of outer space is actually long and daunting. The possibilities include the Inter Agency Debris Coordination (IADC) Committee, the Space Data Association, the Commercial Spaceflight Federation, the International Astronautical Federation (IAF)/the International Academy of Astronautics (IAA), the Committee on Space Research (COSPAR), the International Astronautical Union (IAU), the International Association for the Advancement of Space Safety (IAASS), the International Asteroid Warning Network (IAWN), the Space Mission Planning Advisory Group (SMPAG), and the U. N. Committee on the Peaceful Uses of Outer Space (COPUOS), including its subcommittees and working groups, plus the U. N. Office on Outer Space Affairs as well as the U. N. Office of Disarmament Affairs (UNODA). This is not to mention the other U. N. agencies that are very much involved in space that include at least the International Telecommunication Union (ITU), the World Meteorological Organization (WMO), the International Civil Aviation Organization (ICAO), the U. N. Educational, Scientific, and Cultural Organization (UNESCO), and the World Trade Organization (WTO). It could even be argued that because there are so many formal, semi-formal and informal means of discussing and agreeing to international space policy and law that this very diffusion of the space policy negotiation process gets in the way of agreements actually being reached. It is probably possible to name at least 100 formal and informal mechanisms that exist today to discuss, learn about, debate and possibly work out agreements as to what international outer space policy should be and how it might be improved.

Conclusions

There is at least one thing that is quite clear, that there are today no "space police" to keep order in outer space. The prospect of creating a space police to enforce the peaceful and harmonious use of outer space in the next few decades are somewhere between zero and not on your life.

Even the idea of creating some form of systematic tracking and monitoring system for the regions of space populated only by the regions defined by the Moon, the L-1 Lagrange point and orbits defined by near Earth asteroids is highly unsettling. Indeed, just organizing a system for providing traffic control and monitoring for the protozone seems to be a monumental task.

The B612 foundation has its Sentinel project that involves an infrared telescope that might be used to detect near Earth objects that come within 0.05 astronomical units of Earth. This project is a very ambitious but very specific project to detect the orbits of dangerous space rocks and assess their risk of hitting Earth. It is not a system for monitoring the activity of all those that might seek to approach and establish mining operations on what might potentially be millions of the quite small objects in space quite far from Earth. Detecting space rock orbits versus policing space operations are two quite different things. The cost of building the Sentinel IR space telescope will ultimately be at least $400 million. And this figure does not include its operation or consider that this is for a space instrument with an estimated life of perhaps only a decade.

The cost of monitoring activities for the purpose of space policing would certainly be much more than an order of magnitude greater—perhaps hundreds of times more expensive than scouting for dangerous space rocks. The bottom line is that outer space policing would likely have to focus more narrowly on the control of launch operations when missions left or returned to Earth. This would mean some form of enforcement and licensing process at the time launchers left Earth, brought materials back to Earth, or positioned materials parked in orbit around Earth, the Moon or the L-1 point.

It is one thing to agree to high minded concepts of sharing equitably the wealth of resources in outer space as part of the "global commons" of humankind. It is quite another to make the long-term investment in space transportation systems, robotic space mining systems, power and communications systems to make this a reality. Until there is a viable international mechanism that has the actual intellectual, technical, financial and drive to undertake such a difficult task it will likely not happen as a global enterprise. One of the many difficulties about viewing off-world enterprises as a "global common enterprise" is the lack of a policing and enforcement mechanism.

In the shorter term, the reasonable thing to do is to focus on nearer-term and realizable goals. The first step would be to reach agreement on a way to undertake space traffic management and control for Earth orbits and for the protozone. Such an agreement can at least create a framework and a process to seek to establish space regulatory mechanisms for the longer term future.

References

1. The Outer Space Treaty www.unoosa.org/oosa/SpaceLaw/outerspt.html. Last accessed March 1, 2016.
2. Summary of the Outer Space Treary from the UN Office of Outer Space Affairs. www.unoosa.org/oosa/SpaceLaw/outerspt.htmll. Last accessed March 1, 2016.
3. Agreement Governing the Activities of States on the Moon and Other Celestial Bodies http://www.unoosa.org/pdf/gares/ARES_34_68E.pdf. Last accessed March 1, 2016.
4. "Space Resource Exploration and Utilization Act of 2015." https://www.congress.gov/bill/114th-congress/house-bill/1508/all-info. Last accessed March 1, 2016.
5. Ibid.
6. Ibid.
7. George Robinson, "Space Migration and Colonization," Unpublished article, February 14, 2016.
8. Ibid.

11

Looking Toward a More Hopeful Global Society

Introduction

The new opportunities that space systems and technology now offers to us globally are potentially enormous. As we try to plot a course forward for Earthlings, however, we need to recognize certain truths. Paramount in this regard is that unless we stop literally eating up the limited resources of Earth and start to implement sustainability practices both on the ground and in space, nothing will change. Unless people stop being big consumers and start becoming focused on sustaining and building for future generations, then the riches of outer space will all be for nothing. The new gold rush must be with a purpose. And that purpose is to build a better future.

First and foremost, we need to finally realize that actually, there are not only "human laws" but also "natural laws" of science and the long-term sustainability of life. All of us that inhabit the third rock from the Sun indeed travel on a spaceship. This spaceship travels at about 100,000 km/h around our closest star and completes a solar orbit once a year and a revolution on its axis once every 23 h and 56 min. Incidentally, the other 4 min reflects the fraction of a year represented by the small part of our annual trip around the Sun.

Second, based on our current understanding of physics, astronomy, chemistry and biology, we are protected against various cosmic hazards in what could be seen as almost miraculous serendipity. Earth's atmosphere, as thin as the rind of an apple by analogy, protects us from radiation, provides us oxygen to breathe and water to drink and, for the time being, a livable temperature.

Third, Earth's magnetic field forms the Van Allen Belts. For the most part these rings of magnetically formed ionic matter largely protect us from solar storms and especially periodic blasts of ions that stream out from the Sun.

These are the coronal mass ejections. There are other cosmic hazards that we travelers on spaceship Earth need to worry about as well, such as potentially hazardous asteroids and comets. Fortunately the enormous gravitational fields of the Sun, Jupiter, and even Saturn for the most part help to protect us from being smashed by comets and near-Earth asteroids. To a certain extent one could say that we won the Solar System lottery in terms of getting a location that is just close enough to the Sun to meet our energy needs but not too close, and to have Jupiter's gravitational field to help protect us against comets and asteroids.

Our spaceship planet is in the so-called Goldilocks zone. It is an ideal size, with thermal conditions and radiation level optimal to sustain life. Sustainment of life here on our Goldilocks planet is possible for a very long time if we don't screw it all up via pollution, overuse of natural resources and runaway growth.

This brings us to point four. At the rate that the human population is growing and consuming water, food, and various resources, and with a projected global society that may grow to as large as 12 billion people by 2200, there are limits to growth. We will find ourselves in trouble without some changes to human practices, regardless of what new resources we can garner from outer space. The bottom line is that we must soon stop over-consuming Earth's resources. Our now somewhat cancerous growth of human population across the globe is a real problem. Nigeria's growth at over 6% per annum and India's growth at some 2.5% are literally eating us out of house and home. This is not just adequate food and water, but schools, health care, sewage, jobs, housing, transportation and everything else.

Capturing the bonanza in the skies only helps us if we solve the sustainability issue right here on Earth. Space mining, space processing and manufacturing, space colonies, and even terra-forming of planets are a path forward to a better tomorrow, but only if we conquer climate change and create a sustainable urban lifestyle right here on spaceship Earth.

This adaptive change and a move to sustainable practices can come with new technology, such as solar and wind power, desalinization plants, macro-engineering in space, genetic engineering, and recycling and natural resource sustainability practices.

In the next few decades we can start to systematically reclaim rich natural and energy resources from the oceans and outer space, but this golden age of the future must also involve other changes. Sustainable practices involve population control, expanded global education and health care, more sensible global legal and regulatory systems, and indeed a combination of all of these strategies laid out in a program that can allow, over time, for all to be wealthy and educated. We can empower future generations to have an incredibly sustainable and prosperous future. The bottom line here is that human survival will very likely require a new way to approach the future, embracing new practices based on sustainability and a sensible approach to the great wealth held within what are sometimes called the "global commons." The global commons, as reimagined in this book, include the oceans, the arctic regions, the atmosphere, the protozone, and the major celestial bodies of outer space.

Fifth, the use, exploration and scientific understanding of space involve vital activities that have become a significant part of our human future for the twenty-second century and well beyond. A globally agreed legal and regulatory framework within which all future space activities can be conducted on behalf of humankind will be increasingly important. In fact, without a globally agreed framework for the future utilization of space resources we can be in significant trouble. The lack of such a framework can very likely lead to extreme economic competition, over consumption and abuse of vital resources. The end result will very likely be destructive and quite nonproductive warfare—on Earth and possibly quite soon in outer space as well. With the technology that is now on our doorstep there are riches for everyone if we simply learn how to share and plan for a better and more sustainable future. Two- and three-year olds learn to do this in preschool. Hopefully adults around this small planet can learn to do the same.

If we Homo sapiens end up not only fighting over territory and resources here on Earth but also over the vast reaches and enormous wealth of outer space, it will not only be a great shame but needlessly STUPID as well. We may find we have condemned our species to ultimate doom. And even more ridiculously it will have been for nothing. This is because there are enormous riches to be gotten if we simply learn how to cooperate and share in the vast bounties that are there to harvest. Of all the seven deadly sins, it may well be that the eighth deadly sin, "needless greed," is the deadliest of all.

Realizing the Bounties and Opportunities of Outer Space

We, as humans, need to agree to a new cooperative formula that promotes the development and use of New Space technology, allows entrepreneurial innovations to move us forward, drives down the cost of space transportation and yet also finds a way to meet the needs and aspirations of all Earthlings. This is no easy task. Yet ultimately the riches to be harvested are so great that all humanity can share in this huge win for our species. It is a "win-win" situation if we can just learn to take the longer-term view.

Private enterprise and commercial organizations, developed and developing economies, the worldwide academic community, space agencies, and world political leaders need to find a way to figure this out together. It is actually as simple as learning to "play nice together," and not be too greedy. If we don't insist on always thinking in terms of profits for the short term, but to invest in the future, we can all be enormously wealthy. The first step is to set priorities in seeking a viable and sustainable future.

The great success of Jeff Bezos and Amazon has come from having a long-term plan of development that was not based on short-term profits but having a strategic view of the future. The key to the sustainability of our planet and development of colonies on the Moon and Mars is having a long-term plan of sustainable growth and a global strategic plan for the future that is conceived on the basis of decades and even centuries to come.

A Delicate Formula for Success

There are many ways that this can go very wrong.

Without commercial and technological innovation we will fail. Without a viable way to respond to the needs of all nations to survive and thrive, we will fail. Without facing the need for sustainability we will fail. These challenges actually involve more than the essential need to focus on the growing problem of climate change. Equally important are coping with and developing key systems to deal with issues such as overpopulation, overconsumption, super-urbanization and planetary defense. Without efforts to address all of these issues and more, the advanced economic systems that represent humanity as we exist today may very well fail—or simply fail to thrive.

The current cosmic passageway that opens to long-term human survival is narrow, but the bounties that can be unlocked are enormous. As defined earlier in Chap. 1, it is what Peter Diamandis calls abundance, and the artificial intelligence (AI) Guru Ray Kurzweil calls the golden age of the Singularity.

As we grow to a global economy that is perhaps 90% urban, we are highly dependent on sophisticated infrastructure such as electric power grids, advanced transportation and communications systems, and high-rise offices and dwellings that are subject to potential massive levels of failure—even on a continental or planetary basis. True global "black swan" events, comparable to the killing off the dinosaurs that occurred with the so-called K-T mass extinction event, could be in our future if we do not create new types of planetary response systems quite unlike anything that has ever existed in the past.

This means that we need to be cautious as our planet's vulnerability grows. Our threat level increases each year. This threat rises as we add billions of people to our cities, as we become more and more dependent on advanced infrastructure, and as we become dependent on new types of systems for power, communications, transportation, water and food delivery, and more. We focus always on human threats, terrorism and warfare, but natural and cosmic threats continue to constitute the highest levels of risk.

As all of these changes occur, the sustainability of our species is going to be increasingly at risk. Space resources and space colonies will not save us if we design a future urban society that is increasingly vulnerable to large-scale breakdown.

As these modern urban-based futures unfold, we also become more vulnerable to terrorism, natural disasters, and even technological systems' breakdown. In short, if any of our vital urban infrastructures should fail, the consequences become larger and larger with rapid global population growth and especially with our urban expansion that is rapidly outstripping worldwide population growth.

Fortunately space systems and technology and other reforms can provide us additional protection and resiliency. If we are really clever, it can even bring us increasing global wealth and peace and tranquility. But first we have to stop over-consuming Earth's resources. Second we have to start planning for greater resiliency and redundancy and back-up systems. Third we have to start being a lot better in carrying out longer term planning and developing better governance systems that are fair-minded, farsighted, and forward-looking. Growth of space systems and private space enterprise can be a useful part of this large-scale plan for the future.

Remarkable achievements can be accomplished in coming decades. These include not only clean solar power satellite systems, space mining, material processing, 3D printing and manufacture in space, and other incredible achievements. These might also include observatories and colonies on the Moon and Mars. We might even deploy magnetic space shields to lessen the effects of climate change and ward off solar storms, or perhaps we will create defense systems against asteroid and comet strikes.

The magnetic space shields we create for protecting Earth may also prove to be prototypes to shield Mars from the solar wind, which has stripped off its atmosphere. If such magnetic shields can be proven to work, we could develop a livable and breathable atmosphere on Mars to sustain a whole new civilization on the Red Planet. Thinking even more long term we might ultimately find ways to raise the orbit of Earth so that it can sustain life even after the Sun's heat and radiation begins to expand, making Earth's atmosphere too hot to sustain life. These are just some examples of the long-term potential of future space activities and the future gold rush in the skies. But before all this can happen we need to begin creating a new approach to human strategic planning and global cooperation.

In essence, the right system of global governance for Earth and outer space is actually the key. This would be a system that allows us to raise the potential of all humans to succeed and benefit from science and technology. The technical, economic and creative systems that can allow all humanity to reach its potential are now increasingly being developed to allow a totally new approach to global development. It will require change. There will be a need to redefine wealth accumulation. If we can embrace new forms of cooperation we can realize a cornucopia of benefits. By embracing new forms of cooperation and integrated technical and economic development we can actually unlock longevity, improved health, expanded education, cultural richness and eliminate the perceived value of warfare and human strife.

This may all sound unrealistic and utopian, and many will resist the vision of a bountiful and sustainable world. This is because the idea is does not fit with the thousands if not millions of years of human history up to the present. We are not yet prepared mentally, economically, ethnically, politically, culturally, religiously or intellectually for the time of the Singularity, or the time of abundance. Traditionally ruled political, cultural, and religious societies with totalitarian leaders and indeed many other nation-states will resist such a disruptive change. This new global reality implies a wholesale transition in forms of governance, economics, education and health care systems and even the nature of human employment.

This is a transition that could take a century to unfold. The resistance to this new world order will indeed be powerful—led by religious zealots, powerful economic interests and the ultra-wealthy, leaders of many nation-states, many military interests, and indeed those who simply resist change.

It is the type of economic reality that John Kenneth Galbraith anticipated almost a century too soon when he wrote *The Affluent Society*. Only now are the powerful and indeed highly disruptive technologies needed to make the transition becoming available on a global scale. This abundance will come not only from New Space but also from artificial intelligence (AI) and the application of super-automation to education, health care, industrial production, agriculture and fishing, housing, transportation and communications.

This new world has the potential to make people safer, better educated, healthier, better housed with more food, access to clean and plentiful energy, and eventually protected from natural and cosmic disasters. In addition, this powerful technological transformation will have a side effect of curbing population growth.

This is actually a tall order, and change will take decades. Human history for many millennia has been a struggle for domination of land, vegetation and natural resources. The "wealth of nations" has entailed imposing dominance over people, communities, tribes, nations, and even race. This is because of the struggle to survive against adversity. But through the bounty of technology and the unlimited resources of outer space, the human species could transcend the past and aspire to a greater good, achieving a new level of prosperity while embracing the vital concept of sustainability on Earth and in the skies.

In short, the future challenges we face are more daunting than ever before. In the next 80 years there will be more people to feed, house, educate, receive health care, police, entertain, and help recover from disasters than ever before in the history of humankind. Technology and specifically space technology can help respond to these challenges. Yet without an agreed global pathway to a sustainable future and a solid plan for coping with the limits to growth the long-term prospects for Earthlings are bleak.

It may take decades—maybe even centuries—to devise the technology to create space systems that can colonize the Moon or Mars and to protect us from major cosmic hazards, but the time to start is now. It could take humans millenniums to build capabilities to go to other star systems with any hope of long-term survival. Fortunately we do not have to solve all the challenges at once. Realizing the wealth of resources in the skies can and should help us to start to develop and execute some very long-term plans.

Within 50 years we could build space systems that provide us clean, cheap energy from the Sun and do so 24 h a day. We could create new mega structures in space that could shield us from solar storms and modulate solar heating so that we do not turn into a fireball, like Venus. We could create spaceports at Lagrangian Point 1 (LP-1) that could allow us to go to Mars and mineral- and water-rich asteroids. From this new "way station to the Solar System" we could operate with efficient electrical ion propulsion or solar sails, or other much more efficient systems than the highly explosive chemical bombs we call rocket launchers. We could also create systems to protect us from killer asteroids and comets. Other things that today seem like science fiction, such as space mining, space colonies, space processing and manufacture, and more, could also be a part of tomorrow's reality. As a critical first step we need to create a systematic global space agreement that creates the right incentives for New Space entrepreneurs, technologists, national governments and populations to move forward to realize the value of the great riches in the skies.

Some of these activities such as space mining, solar power satellites, and mega structures plus innovative space systems will take investment and time to realize, but the bonanzas in the skies are actually enormous. Thomas Jefferson's buying of the Louisiana Purchase or Seward's Icebox (namely Alaska) were seen in their day as financial extravagances but today are recognized as the world's shrewdest investments. These are nothing in comparison to the trillions and quadrillions of tons of riches in the skies. Mega projects in space to shield us from powerful solar storms via newly constructed "artificial Van Allen Belts" may sound like exotic and expensive ventures. But such structures are not a waste of money or extravagant if they turn out to be essential planetary defense systems that are needed to sustain modern society and preserve the human race and modern society as we know it.

The short way of saying this is that the coming gold rush in the skies is based on a number of critical success factors (CSFs). In short, we need to plan for and implement some truly unique innovations that will be unique to the twenty-first century. Oddly enough the most important of these factors do not hinge on familiar topics such as more and better jobs or lower taxes or less governmental waste or better security measures against terrorist activities. No, humanity must recognize in the age of the Singularity and the new opportunities in the skies that these are actually problems of past centuries if we just allow ourselves to look to new opportunities in the skies that can bring us totally new wealth. That is at least half of the equation. The other half is to recognize that we must also invest in New Space systems to provide ultimate protection against dangers from the skies.

And the most essential key to the future is global agreement that leads us to a sustainable and survivable world. This cannot be based on vague platitudes. No, the key would be economic formulas that are based on wins for all the key players.

To Make Humans Great Again

Here we steal a catch word from Donald Trump. The idea, though, is not to make the USA great again but to transform our global economy and its worldwide human services to make all Earthlings great. And one of the key steps to achieving this objective would be to ensure the peaceful use of outer space and to find ways to do this so that it would lead to tangible benefits for all humankind.

There must be a caveat added here. And that caveat would be that those that provide the capital and the technology to exploit the benefits of outer space must also receive recompense for their efforts as well.

Space dividends in this grand plan would benefit space commerce and entrepreneurs, spacefaring nations and non-spacefaring nations, technically advanced nations and non-technically advanced nations. The key would be new partnerships that would be forged from a new spirit of cooperation. Too often efforts in this direction have been sought to be imposed from the top down through the Outer Space Treaty and especially the Moon Treaty. The key would be to find new mechanisms to build these space-partnerships from the ground up. The future dividends of space are sufficiently great that all could and should share in the benefits.

If we are truly innovative and forward thinking we can truly invent a better and richer future for all humans. We can afford to be generous and share in the bounty if we simply agree to forego needless greed. Sharing in the sustainable riches of the gold rush in the skies is possible if we can recognize that infinite resources can change the economic equations of the past and open up a bountiful future.

Key elements of these new types of global partnerships already exist. There are already institutions such as the Singularity University and the International Space University that are geared to find new ways to solve our global economic, social and technological problems.

Young people are brought together from all over the world to learn about the potential of the future and to find new ways to invent a better future. Here they can work across national lines of division, innovate and conceive of new technology and systems that transcend the old ways of looking at how to do thing. The Singularity University has a very different way of looking at the future. This unique body brings young people together to learn new skills and work together, but it is much more than that. In this case they are not just focused on space but on artificial intelligence, information technology and networking, robotics and other innovative systems. Their assignment is to create new enterprises that in a decade can have a positive impact on millions of people. They are not looking to learning a new craft or creating a new company to generate personal wealth but to transform the future.

These new institutions, just like Google, are not seeking to improve a product or service by 1% or 2% but by orders of magnitude. They are seeking disruptive new technologies that can change the world and make the old way of doing things obsolete. The idea arises from the so-called singularity concept as envisioned by Ray Kurzweil, plus the concept of abundance as envisioned by Peter Diamandis, and then amplified by the game-changing approaches to the future as thought of by Pete Worden, former head of NASA Ames. These three innovative thinkers started the Singularity University—not to change educational options but to allow the world to escape from its past.

We humans have some very big problems, such as climate change, excessive population growth, overconsumption of key vital resources, ethnic, cultural, and religious rivalries, and more. Today there are disparities in many areas of the world, disparities that include education, health care, housing, access to potable water, and more.

Clearly there are enormous problems around the globe, and it is correct to address ways to expand education, health care, as well as cope with the challenges of super-urbanization and to find ways to bring peace to regions enmeshed in warfare. Unless we solve these problems it could lead to the failure of the human enterprise as we know it today. What is overlooked in the quests to solve these Earth-based problems is how to deploy New Space systems—linked to ground-based technology—that can truly can help to solve problems here on our small planet we call Earth.

In the next hour of "Super Month" time we can fix many of humanity's current day problems and bring wealth and prosperity to all. Here's how.

A Ten Point Program to Enrich the Human Experience and Unlock the Potential of Outer Space

We are not being naïve. The way forward is not easy. Below is a simple, but very difficult to execute, ten-point program. If we could follow it, the result would be to let humanity reach its potential and realize the riches that can be released from the heavens above, plus a more effective way to use the natural riches of our planet here on Earth. The truth is that there need be no poverty. Rather there is a profound disparity in how we organize our political, economic, social, and cultural systems. The result is warfare, prejudice, racism and needless suffering. Humanity is now capable of curing disease, colonizing celestial bodies, and overcoming poverty and needless aggression. It is time that we do just that. Space systems and related development are just tools that allow us to achieve worldwide prosperity if we simply stop doing thoughtless and unproductive things.

Here is the ten point program that could make the world safer, more prosperous, and help us suffer much less strife and warfare.

1. Seek agreement on global population control and associated incentives as a means to stop eating our planet alive.
2. Move to adopt sustainability as a strategy to create and achieve new wealth.

3. Encourage the world's space agencies to prioritize planetary protection programs—including space shields to preserve modern infrastructure.
4. Recognize that urban sprawl is bad, urban density is good, while super urbanization and mega cities are bad again.
5. Refocus efforts on developing systems to address priority needs related to climate change, potable water and U. N. goals for sustainable development.
6. Deploy new space- and ground-based infrastructure for education and health care.
7. Recognize the unrealized potential of the coming singularity and the new abundance.
8. Adopt a new system of laws and regulation for the twenty-first century for global cooperation in space and on the planet.
9. Redefine and re-incentivize exploitation of "the global commons" for the twenty-first century based on sustainability and equitable sharing.
10. Recognize that humanity is at the turning point. It is time to redefine the nature of jobs, human toil, wealth, and recalibrate economic and legal systems.

Key to successfully adjusting to our new future is finding creative ways to share the cosmic commons that are vital to human survival. Thus this chapter reviews both the challenges and the possible new regulatory and legal standards, and revised practices that could allow us to realize the unlimited opportunities of a vast space that beckons us to the future. These "commons" of outer space and here on Earth can help unlock humanity's unlimited potential or be the pathway to strife and warfare.

Too often we look to the future through the rearview mirror of history. We cling desperately to the leavings of history and its consistent failures to find societal knowledge, peace, prosperity and global tranquility. We are wedded to the devil we know, rather than the unlimited potential of a future that is rich with unrealized bounties of knowledge and creative endeavors to unlock new riches here on Earth and in outer space.

The following ten guidelines that could unlock magnificent riches of the future depend on a totally fresh perspective. They hinge on realizing that smart machines, broadband systems, space systems and technology, and new ways of looking to the future can redefine human potential, redefine wealth and make the old way of doing almost everything obsolete. A new system of Global Governance for Outer Space will not only unlock enormous potential in the cosmos but on planet Earth as well.

Here are ten possible steps to a more rational way to use the natural riches of our planet.

1. STRUCTURING GLOBAL POPULATION CONTROL AS A MEANS TO STOP EATING OUR PLANET ALIVE. The most remarkable period of economic development ever recorded on Planet Earth has been achieved by the People's Republic of China. Its average Gross Domestic Product per capita in the past 40 years has increased by a factor of 16. No other country has ever made similar economic progress. A variety of factors, including global trade opportunities, economic stimulation policies, and international investment have combined to spur this remarkable growth in prosperity. Today there are more billionaires in China than in the United States.

One of the most significant factors that led to the economic miracle was an effort to minimize population growth and to encourage urban industrialization. In contrast, those countries projected to continue rapid population expansion such as Nigeria, India, Bangladesh, Uganda, Democratic Republic of Congo, Ethiopia, Pakistan and Tanzania, will face economic stagnation and see poverty grow. There will also be a voracious appetite for housing, water supply, food, education and health care services, energy and natural resources that will be a constant struggle to meet. Failure to improve quality of life is almost guaranteed. India is expected to grow by 450 million, Nigeria by 270,000, and Pakistan some 140,000 in the next four decades. In contrast China may shrink by as much as 400 million. China will prosper, and their success will likely be a key factor in coping with climate change.

In short, curtailing population growth is not only the key to improved services, better education and health care, less pollution, better transportation systems, and effective ways to combat climate change. There is a huge global database that supports the thesis that the prime pathway to sustainability of human life on planet Earth is population containment and zero population growth. Such reductions in population growth is the first step forward to personal prosperity, improved citizen services, a better educated populace, balanced diets, viable strategies for coping with climate change and sustainable and continued economic growth and prosperity.

There is only so much water, so much iron, copper, wood, and oil on our finite planet. Unless effective population growth is achieved in the twenty-first century, the limits of growth for our small world will be reached by the end of the twenty-first century. The first step to a new golden age for humans is keeping human population growth in check. If we can recycle the bounteous resources we have on Earth efficiently and economically, then the vast resources that we can import from outer space become big cosmic bonuses. These bonuses we can invest in projects to create systems to insure our long-term survival and future prosperity. With the amazing New Space systems and technologies—combined with innovations right here on Earth—we can make everyone healthy and long-lived and prosperous in every imaginable way. In short we can create systems to save ourselves from cosmic hazards, create clean and low-cost energy systems that can power amazing new infrastructure and provide the bounty we need to create amazing new communications, information technology, transportation and robotic systems that were once thought of as the fantasies of Buck Rogers science fiction stories.

2. SUSTAINABILITY AS A STRATEGY TO ACHIEVE WEALTH. There are those that dismiss the ideas of recycling, sustainability, or green systems as the misguided activities of "do-gooder tree-huggers." It must be recognized that these are not idealistic dreams by green-obsessed environmentalists, but likely to be the very engine of twenty-first century economic growth. The bottom line is that sustainability and New Space initiatives are perhaps the world's most significant job expansion opportunity. The race is already on. China has its foot on the accelerator to become the largest supplier of photovoltaic cells, solar array systems, state of the art high speed trains, advanced battery systems, etc. Germany leads in wind energy systems. Japan is leading in development of solar power satellite designs.

The future today is not what it used to be. Today it is sustainable energy, transportation, communications and virtual reality (VR) telecommuting systems. It is even advanced building materials systems, housing built by 3-D printers, and "smart" building systems designs. These state of the art sustainable products are not being promoted by bleeding heart liberal environmentalists but are being developed by today's leading figures, such as Tim Cook of Apple, Mark Zuckerberg of FaceBook, Elon Musk of Tesla, and other forward-looking smart industrialists around the world.

3. PLANETARY PROTECTION TO PRESERVE MODERN INFRASTRUCTURE AND AVOID GLOBAL DESTRUCTION. One of the keys to the future of human survival is for the world's space agencies to finally recognize that their top strategic objective is to deploy space systems that would save Earth from massive destruction. You would think that this is self-evident, but it unfortunately isn't. The Administrator of NASA, along with the leaders of other space agencies, believes that their job is to explore the Solar System and create scientific satellites to better understand the cosmos. There is absolutely nothing wrong with such objectives, but they are failing to face up to the reality of what they are overlooking. Not once have the space agency administrators paused to say: "Duh, if we don't have a world with a world economy we are out of business." Certainly if there are no people or financial resources left, space research and exploration goes away.

For too long, there has been the assumption that cosmic hazards are so remote and so difficult to cope with that the only thing to do is to ignore them. Finally the U. N. General Assembly has agreed to set up the International Asteroid Warning Network (IAWN), and the Space Mission Planning Advisory Group (SMPAG) and NASA repurposed the WISE Infrared Telescope to hunt for near Earth objects that might be potentially hazardous. However, NASA, the European Space Agency (ESA), the Japanese Space Agency (JAXA), the Indian Space Research Organization (ISRO), the Chinese National Space Agency, and the Russian Space Agency (ROSCOSMOS) spent on the order of one-tenth of 1 % of their budgets on cosmic hazard detection and planetary defense.

This is a huge dereliction of their duty to save the human enterprise. One of the keys to achieving the gold rush in the skies is for the space agencies to start doing their jobs. There are at least three top priorities.

The first is to create a Sun shield against solar coronal mass ejections that are going to become increasingly lethal as the shift of Earth's magnetic poles erode the natural protection offered by the Van Allen Belts. (A Lloyd's of London study has projected $2.7 trillion dollars in losses from a big solar hit. That's "trillions" of dollars.) We are talking about the loss of satellite networks, loss of electrical power grids, and likely huge loss of human life after such a big hit occurs. The solar threat that is now building as the magnetic poles shift will increase every year for decades to come. The time has passed for space agencies only to say this is a growing cosmic danger, but we have absolutely no ideas about how to build a protective shield against a huge solar storm hit. Surely we are smarter than the dinosaurs. Anyway this should be job one.

The second priority is to develop better tools to identify asteroid and comet threats and in particular using infrared telescopes that can spot city killer asteroids that are much smaller than the 230-m space rocks they are now looking for. Even 35-m asteroids at the right relative speed could wipe out a mega-city like New York, Beijing, Mumbai, Rio, Tokyo or London. In parallel we need to develop the tools to divert the threat such as a new orbit into the Sun.

The third priority is a new and effective means to decrease orbital space debris and develop the technology to get the most dangerous "space junk" out of Earth orbit. This dangerous debris threatens our very access to space in the future.

The amazing thing is that if the various space agencies around the world were to devote something like 5 % of their budgets to these three tasks they could make significant headway in all three of these key elements of a planetary defense program. For too long there has been minimal progress made against these dangers because of what is a stupendous case of cosmic myopia. The time of change is now. It is time for a clear-cut change in the top strategic goals for all the world's space agencies.

Save Earth first! Explore and research outer space second.

4. THE UNREALIZED POTENTIAL OF MEGA-STRUCTURES AND INTELLECTUAL INFRASTRUCTURE IN SPACE. It is possible that some space scientists and experts will say, "But the Sun's power and processes are much too vast for human tools to change its behavior." The answer is that one does not have to change the Sun's behavior. We only need to create a solar shield, most likely at the L-1 Lagrange Point some 1.5 million km away from Earth—at the "balance point" between the Sun's and Earth's gravitational forces. Since the protective shield built to divert mega-sonic ions from the Sun would be positioned quite distant from Earth (i.e., four times further away than the Moon), the magnetic shielding system mounted on balloon-like structures would not have to be too large. This system could also be built to modulate solar radiation to slow climate change. And it could also serve as a way station to transition from chemically powered rockets to higher efficiency and lower cost electric or ion propulsion systems. Finally it might even be designed to convert solar radiation to clean power that could be beamed back to Earth 24 h a day.

This is just one concept of a space structure that could be built to provide new capabilities back on Earth. The idea of a radiation and solar CME ionic shield is the initial design feature, but solar power and transportation way station features might be added without excessive cost. If such a new structure could be designed and built for less than the cost of the International Space Station, it would seem to pass the "reasonable" test. The ISS really was no more than a space research station; in contrast this solar shield and transport terminal would protect Earth from trillions of dollars of devastating losses, perhaps save millions of lives, provide a constant supply of clean energy from space, and serve as a cost-efficient way station to the rest of the Solar System. And this is just one idea about how humans could reshape the structure of the inner Solar System to make it more usable and to help protect Earth from future catastrophic losses [1].

5. URBAN SPRAWL IS BAD, URBAN DENSITY IS GOOD, SUPER URBANIZATION IS BAD AGAIN. For many years urban planners have sung a consistent song about the dangers, wastes and polluting effects of urban sprawl. The thought is that urban density can make for more efficient transportation, water, handling of sewage, communications, power, and storm sewer systems, and that housing, offices, schools, medical facilities, and shopping areas are also more effective and cost efficient. The problem is that in almost every system that can be devised there comes a point of diminishing returns. This does seem to be true

about everything. A building that grows too tall is ultimately overtaken with all of its internal volume devoted to elevator shafts or escalators. A building that is too tall or massive is very difficult to cope with if there is a fire, a power failure that shuts off lights and elevators, a failure of heating, ventilation, air conditioning, or perhaps worst of all, a terrorist attack. Today there are over 25 mega-cities of over 10 million people, and the number of people in the world as of 2100 is projected by the United Nations to be as high as 12 billion.

Some have despaired about the future of our overpopulated world that is becoming polluted, its national resources and potable water supplies used up, and in danger of overheating due to climate change. They suggest that only creating colonies on the Moon or Mars can help sustain the human race. This seems overly pessimistic. The key is for humans to solve the sustainability puzzle for both Earth and outer space colonies as one large integrated problem.

On Earth and in outer space the formula seems to be exactly the same. These sustainability rules are actually fairly simple: (1) Develop systems to sustain zero population growth at a level that sustains prosperity, regeneration of natural resource supply and potable water, clean energy supply, and genetic diversity. (2) Create living, working, recreation and culture standards in dense urban environments that achieves economic, energy, and transportation efficiency, and proximate food supply. This means avoiding super density where there is loss of community spirit and involvement. Such super density also creates major problems of response to natural or human-made disasters or terrorist attack. (3) Finally re-focus on new approaches to urban planning devoted to research to achieve more effective use of artificial intelligence and automation so as to create full employment. This new thrust would be toward achieving universal education, health care and employment systems geared to a sustainable world and coping with the global strife associated with cultural, racial, linguistic or social conflicts.

6. SPACE AND GROUND-BASED INFRASTRUCTURE FOR EDUCATION AND HEALTH CARE. A significant part of this new world and space economic and governance system would be to create systems geared to providing universal education, health care and community service. Space-based and terrestrial information technology (IT), artificial intelligence, robotics and communications systems will become more and more essential to achieving universal global education and health care networks that reach every person on Earth and also in space-based living quarters and eventually space colonies. Today 27% of the U. S. gross domestic product is devoted to education, training and health care. A full deployment of "smart technologies" can expand education and health care services to everyone, but at a small fraction of their previous

costs. These systems, through space-based satellite networks, can reach every person in the world. One of the consequences of these changes will be a loss of linguistic, cultural and social diversity, but this is an almost unavoidable consequence of a highly educated and Internet-connected world.

7. THE UNREALIZED POTENTIAL OF THE COMING SINGULARITY AND THE NEW ABUNDANCE. The last 200 years of global history has shown a consistent pattern of change. Increasingly employment has shifted away from the primary industries of farming and mining. Around 2–3 % of the jobs in economically developed societies are in these sectors that are now highly automated for plowing, watering, adding of fertilizer and harvesting when it comes to farming, and mining is also more often done by machine. Industrial manufacturing is likewise being automated not only for machines such as cars but even computers, medical and scientific instruments, cell phones and robotics. The biggest employment area in the service sector is now education and health care, and this too is being rapidly "automated" through the use of artificial intelligence and programmed learning systems.

Ray Kurzweil, the father of SIRI (of smart phone fame) and general-all-round AI guru, has suggested that the so-called Singularity, where robotic machines with "humanoid-like brains" will be on assembly lines, will be available for mass consumption within homes and offices in the next few decades. Not only will such technology free up workers in farming, manufacturing, education, health care, marketing, transportation, banking and insurance, but this will spread to virtually all other service industries (with only a few possible exceptions, such as prostitution). In the space industry, it will play an ever increasing role in exploration, scientific investigation, and certainly will play a key role in space transportation (i.e., automated spaceships and spaceplanes).

Perhaps most significantly of all the artificially intelligent "thinking machines" will be able to invent new products, services, and change even our understanding of chemistry, physics, astronomy, geology, biology, human anatomy and pathology, genetics, electrical engineering, synthetic materials, computer sciences, and information technology and communications, and even what is artificial intelligence. Machine-driven innovations may well become much more rapid than human-based research initiatives. This is to say, the world will change forever once the so-called Singularity begins to truly change the worlds of education, medical services and research, human employment, inventions and disruptive technologies.

Things that might be anticipated are "quantum dot" solar cells that can give us clean and cost-effective energy from the Sun, "thinking" robots to carry out household chores, maid services in hotels, nursing duties, virtually every basic service now providing the bulk of employment in developed economies, whether it is fast food workers, banking clerks, accountants, couriers, flight attendants, and on and on. The long and short of it is that the Singularity will mean a transformation in what it means to be employed. Everything will change—capitalist economics, value chains in any industry or service enterprise one can imagine, even the concept of wealth. Neither the global economy here on Planet Earth or the world of space research and applications is ready for what this new world will bring in terms of economic turmoil or potential unrest. The emphasis will change from the individual to the community in this new world of super automation and accelerated innovation and change.

The stakeholders with the biggest share of global wealth and control of industry and private wealth will seek to resist this massive economic, social and technological upheaval. Likewise those that are fearful of technological advance, and wish to retain traditional or totalitarian forms of leadership and governance, will seek to forestall not only the technology but also the cultural, social and religious changes that are implied with this sweeping evolution of human history. But massive and revolutionary technological change will be hard to resist. In the space arena, the possibilities of rapid change to support space mining, solar power satellites, space colonization or even terraforming of the Moon and Mars may all accelerate rapidly. In short, the shift to a space-based economy of the future will be driven by Earth-based changes known most succinctly as the Singularity.

What was uneconomic or unthinkable just decades ago, may suddenly become technologically and economically feasible at a very accelerated pace. Ultimately there is a clash on a global scale between Western technological advance, summed up as the coming Singularity, and cultural resisters represented by some terrorist groups. It is ultimately not religious but a cultural and technological conflict that will ultimately determine the governance systems that will prevail in the centuries ahead.

8. A New System of Laws and Regulation for the Twenty-first Century for Global Cooperation in Space and on the Planet. This massive social, economic and technological change triggered by the Singularity will perhaps have its greatest impact on what is generally called governance. In this context we are talking about the laws, regulations, ownership and mechanisms that control human activities around the world. Old economic incentives will no longer work—or certainly not work as well as they once did.

9. Redefining the Global Commons for the Twenty-first Century— New Systems, Ways and Means to Exploit the Oceans, the Arctic Regions, the Protozone and Outer Space. There are many things that are in conflict in the world today. There are those that promote technological

advance and the coming Singularity, or what Peter Diamandis characterizes as the coming abundance. These are the innovators that are creating the disruptive technology that will lead to change in the world of space, information, artificial intelligence, robotics and electronic systems.

Then there are those that advocate a return to traditional values of fundamentalist beliefs and hegemony of male elders and absolutely no change in traditionalist beliefs. Trapped in the middle are the existing major stakeholders in today's economic systems based on hydrocarbon fuels, multi-national banking and insurance corporations, and the so-called "one percent" that control much of the world's wealth. These wealthy leaders would like to see a very slow movement toward innovation. They are made nervous by a rapidly changing social environment and particularly by the prospect of an emerging new world driven by millennials and calls for a new international legal system that focuses on a global commons or suggests that generations that are yet unborn might have legal rights. These are the world's most wealthy capitalists who do not wish to see the *status quo* disturbed. They become particularly nervous when international legal experts talk of a governance regime that extends to all the resources and opportunities presented by the oceans, Antarctica and the polar regions, outer space and the stratosphere (or the protozone), where all are owned as part of a "common world human heritage." In this brave new world all inhabitants on our planet—namely spaceship Earth—have a joint stake. The choices seem to boil down to three options. First there is a "technocracy" that supports redefining work, wealth, social values, and a move toward global governance on Earth and in outer space. Second there are those that support the global *status quo,* where a capitalist economy drives the industrial machines toward more consumption and ever more rapid throughput of product and mounting pollution. Then there are those that support governance based on fundamentalist forms of religiously founded leadership.

There are several keys here. One key is the clear acceptance or rejection of the idea of a global commons. This viewpoint supports common control of resources not under the control of a single nation and that may involve vital assets for our human future. Another key is the perceived growing need for some form of a global governance system to cope with key common issues such as climate change and planetary defense. The question posed here is whether or not we ultimately accept the reality that our world is indeed a 6 sextillion metric ton mud ball that is hurtling through space with the Sun at about 107,000 km/h (66,000 miles/h). This requires us to ultimately accept what we have now known for centuries—that the fate of all humans and indeed all flora and fauna that inhabit our tiny planet is shared by one and all, and our lives are quite dependent on the Sun and cosmic events.

There are the other common issues hanging fire as well. These include whether we can agree to undertake a true joint and common approach to the cleaning up of the oceans and the atmosphere as well. A close subsidiary question is whether we can figure out an equitable joint use of the polar regions. Another question is how to regulate and undertake the joint exploitation and fair use of the resources represented by the stratosphere and near Earth orbit, the protozone. Embedded in that question is how to establish and pay for a proper system of space traffic management and control. Finally we come to the largest governance question of all. And that is how we can use outer space to the common benefit of all humankind? These are complicated questions related to economic investments, development of new technology and assuming different levels of risks are necessary to undertake needed action. Some nations and companies are more likely to have the technology, the investment money and the will to make the needed future changes happen. Without these actions by a few, the needed future benefits will not come about.

Equity seems to suggest that those who make the necessary investment, take the needed risks and develop the new technology should somehow be properly rewarded. Still, it is far from clear as to what the equitable path forward should be. Currently the U. S. Commercial Space Act of 2015 suggests one way forward under one set of rules. But the Outer Space Treaty and the Moon Agreement is viewed by others as providing a different set of answers.

Past governance systems that have tried to utilize new resources and develop new capabilities include socialism, communism, free-market capitalism, democratic socialism, representative democracy, and even other names. All have found varying degrees of success and failure. All attempts at creating global forms of cooperation and governance, such as the League of Nations, the United Nations, even the European Union, have had some degree of success—and failure. Instruments such as the Law of the Sea and the Outer Space Treaty have likewise proved less than perfect arrangements. But it is critical that we keep on trying to find better answers. Efforts to share in the key tasks ahead to save Earth, to exploit the riches of outer space, and to find ways to reward innovation will not be easy. Finding the way to encourage the pathfinders while also aiding those most in need ultimately seems a very humanitarian and worthwhile goal to set for the twenty-first century world. It is difficult to find the right incentives to reward those that wish to conquer the cosmos and also save our planet and those back here on Earth that rightfully feel they should share in the bounty.

10. Humanity at the Turning Point—Redefining Jobs, Wealth and Economic and Legal Systems. There is no time like the present to recognize that cosmic time for humanity is not only speeding up but accelerating

to warp speed. Biological, information and space technology are exploding. Increasing global population and super density of cities, super automation and artificial intelligence, the redefinition of work and wealth, climate change dangers and planetary challenges—all are coming to the boil. And these are not happening as a nice steady progression of events but all at once. When one lives in turbulent times it is hard to calibrate the pace and suddenness of change, but we are entering a funnel of change like none other, or like that of the 4 million year period that has seen the emergence of the southern ape man on Planet Earth.

Conclusions

Both dangers and opportunities are reaching their maximum levels against the ticking of a cosmic clock of change. New legal systems, new economic incentives, new technology and even new planetary defense mechanisms will be needed all at once. There is indeed a new urgency to finding answers on the fulcrum of sustainability, human equity, and shared striving for a better tomorrow. Our biggest problem is realizing that these challenges, opportunities and dangers are all wrapped together in a crucible of change, innovation, and transition to a totally new world—a world different than we have ever faced before. The question is whether we are up to the challenge. Can we bend our economic, political, social and cultural wills to find a better way forward? We will have to work together as never before to put survival of the human species ahead of personal wealth. We will have to let smart machines help us find new solutions in education, health care, manufacturing, urban planning and even birth control to get us through the next five decades of incredible change. The absolute turmoil of these decisions is reflected in the stresses and strains that are showing up in the European Union, the presidential election in the United States and the economic upheavals being seen in China. Truly humanity is at the turning point.

Let's hope we can make the right choices in governance and make the transitional changes necessary to face the brave new world.

Reference

1. Joseph N. Pelton, "Let's Build a Megastructure in Space to Save Earth," *Room Space Journal,* Summer, 2016.

Appendix: Current Status of the U. S. Commercial Space Launch Competitiveness Act, Public Law 114-90, as of June 2016

In December 2015 the U. S. Congress passed Public Law 114-90 that was signed into law by President Obama. Part of the requirements of that act was that the President was required to recommend a process whereby the U.S. would fulfill the requirements of the Outer Space Treaty to provide oversight of activities carried out by U.S. entities in space. The following report from the White House Office of Science and Technology Policy fulfills that obligation to Congress. This report is included because it characterizes the many U.S. initiatives to carry out commercial space activities that are now proposed and are considered actively pending.

EXECUTIVE OFFICE OF THE PRESIDENT
OFFICE OF SCIENCE AND TECHNOLOGY POLICY
WASHINGTON, D.C. 20502

April 4, 2016
Dear Chairman Thune and Chairman Smith:

This letter is submitted in fulfillment of a reporting requirement contained in the U. S Commercial Space Launch Competitiveness Act (Public Law 114-90, herein referred to as "the Act"), signed into law November, 25th, 2015. In addition to updating and expanding Title 51, United States Code, the Act requires the development of a number of reports on commercial space matters. Section 108, Space Authority, provides:

(a) IN GENERAL—Not later than 120 days after the date of enactment of this Act, the Director of the Office of Science and Technology Policy, in consultation with the Secretary of State, the Secretary of Transportation, the Administrator of the National Aeronautics and Space Administration, the heads of other relevant Federal agencies, and the commercial space sector, shall

1. Assess current, and proposed near-term, commercial non-governmental activities conducted in space;
2. Identify appropriate authorization and supervision authorities for the activities described in paragraph (1);
3. Recommend an authorization and supervision approach that would prioritize safety, utilize existing authorities, minimize burdens to the industry, promote the U.S. commercial space sector, and meet the United States obligations under international treaties; and
4. Submit to the Committee on Commerce, Science, and Transportation of the Senate and the Committee on Science, Space, and Technology of the House of Representatives a report on the activities described in paragraphs (1), (2), and (3).

(b) EXCEPTION—Nothing in this section shall apply to the activities of the ISS national laboratory as described in section 504 of the National Aeronautics and Space Administration Authorization Act of 2010 (42 U.S.C. 18354), including any research or development projects utilizing the ISS national laboratory.

(1) Assess current and proposed near-term, commercial non-governmental activities conducted in space;

United States companies presently engage in an array of space activities, such as launch services, satellite communications, and remote sensing, which are regulated (1) by the Secretary of Transportation, as delegated to the Administrator of the Federal Aviation Administration under Chapter 509 of Title 51; (2) by the Federal Communications Commission under the Communications Act of 1934 (47 U.S.C. 151 et seq.); and (3) by the Secretary of Commerce, as delegated to the Administrator of the National Oceanic and Atmospheric Administration under Chapter 601 of Title 51. The Administration understands Congressional interest in this report is not on the aforementioned current activities, but instead on newly contemplated commercial space activities.

A number of American companies that are investing in the development of innovative, unprecedented space activities have indicated that their proposed activities in space could begin in as early as 1 year or might not begin for a decade or more. This section broadly describes three categories of unprecedented commercial space activities planned by American companies.

Private Missions Beyond Earth's Orbit

- Multiple American companies have announced plans for commercial missions to the Moon, including transportation of commercial payloads to the lunar surface. One such company has indicated that it has a launch contract for a technology demonstration mission to the Moon, which would involve maneuvers on the lunar surface.
- One American company has announced plans for commercial missions to Mars in the near future.
- One American company has announced plans to operate a commercial lunar habitat.

New On-Orbit Activities

Several American companies have announced plans for new on-orbit activities, with start-up time horizons ranging from 1 year to decades, including:

- End-of-life extension modules, which attach to a satellite to aid in station-keeping or transfer to a graveyard orbit;
- Satellite repair utilizing robotic arms;
- Satellite refueling utilizing fuels launched from Earth;
- Satellite refueling utilizing fuels derived from space resources; and
- Commercial orbital habitats.

Space Resource Utilization

American companies have announced long term plans to extract resources, such as rare-earth elements from the Moon or asteroids, for use on Earth or in space as a means of supporting deeper exploration and a longer-term human presence in space.

(2) Identify appropriate authorization and supervision authorities for the activities described in paragraph (1) ["current and proposed near-term, commercial non-governmental activities conducted in space"];

In addition to implementing U.S. international obligations, the existing arrangements for authorization and supervision of non-governmental activities in outer space are designed to serve a range of public policy interests, including public safety, safety of property, national security, and foreign policy. The United States has a legal obligation under Article VI of the 1967 Treaty on Principles Governing the Activities of States in the Exploration and Use of Outer Space, including the Moon and Other Celestial Bodies ("Outer Space Treaty"), as follows:

States Parties to the Treaty shall bear international responsibility for national activities in outer space, including the Moon and other celestial bodies, whether such activities are carried on by governmental agencies or by non-governmental entities, and for assuring that national activities are carried out in conformity with the provisions set forth in the present Treaty. The activities of non-governmental entities in outer space, including the Moon and other celestial bodies, shall require authorization and continuing supervision by the appropriate State Party to the Treaty.

Article VI arose from one of the more contentious issues in the negotiations leading to the Outer Space Treaty. The Soviet Union strongly favored a formulation that would have restricted space activities to governments. The United States, whose companies had plans for privately operated telecommunications satellites, urged a formulation preserving the possibility of non-governmental space activities. Article VI codifies the bargain that resolved the impasse.

Many space-faring States discharge this treaty obligation through a more general licensing framework for non-governmental space activities. The United Kingdom's Outer Space Act of 1986, for example, establishes a single licensing process for all space activities conducted by UK nationals (with the exception of spectrum-related issues), and ensures conformity with the provisions of the Outer Space Treaty, and other public interests such as national security, through license conditions. Likewise, the United States utilizes license conditions to implement its international obligations and to safeguard public interests, but utilizes separate frameworks for licensing launch and reentry, remote sensing, and communications. Although these frameworks have served the United States well by addressing the commercial space activities to date, they do not, by themselves, provide clear avenues through which the United States Government can fulfill its Article VI obligations in relation to the newly contemplated commercial space activities described in Section 1.

The unprecedented commercial space activities described in Section 1 of this report, such as activities on the Moon and other celestial bodies and utilization of space resources, implicate the provisions of the Outer Space Treaty in ways not clearly addressed by the existing licensing frameworks. While existing licensing frameworks provide clear means to address certain aspects of these activities, they do not, by themselves, provide the United States Government with a straightforward means to fulfill its treaty obligation to ensure the conformity of these activities with the provisions of the Outer Space Treaty.

The Administration is actively pursuing mechanisms, including the legislative proposal described in Section 3, to enable the Government to authorize innovative new space activities by U.S. companies consistent with cornerstone treaty responsibilities and obligations.

(3) Recommend an authorization and supervision approach that would prioritize safety, utilize existing authorities, minimize burdens to the industry, promote the U. S. commercial space sector, and meet the United States obligations under international treaties; and

The economic vitality of the American space industry is best served with a clear and predictable oversight process that ensures access to space and imposes minimal burdens on the industry. The Administration supports a narrowly tailored authorization process for newly contemplated commercial space activities, with only such conditions as are necessary for compliance with the United States' international obligations, foreign policy and national security interests, and protection of United States Government uses of outer space.

Through months of consultations among Federal departments and agencies and with the commercial space industry, this Office developed a legislative proposal for a "Mission Authorization" framework, which is appended to this report.

Through the Mission Authorization proposal, the Administration does not seek to establish a comprehensive regulatory framework for the type of outer space activities described in Section 1. At this early stage in the development of these activities, consisting primarily of experimental technology development and demonstration, the Administration believes it would be premature to establish a comprehensive regulatory framework mirroring those for mature commercial space activities, such as launch services. Instead, the proposed legislation is intended to establish a process no more burdensome than is necessary to enable the United States Government to authorize these pioneering space activities in conformity with its treaty obligations, and to safeguard core public interests, such as national security. By providing a clear path for authorization and supervision of new space activities, the legislation would encourage investment in those activities and foster and promote a robust domestic commercial space industry.

The Mission Authorization proposal is closely modeled on the FAA's Payload Review process, in that the FAA would coordinate an interagency process in which designated agencies would review a proposed mission in relation to specified government interests, with only such conditions as necessary for fulfillment of those government interests. For example, the Department of State would be responsible for reviewing proposed missions for consistency with the Outer Space Treaty, and would recommend authorization conditions only as necessary to ensure conformity with the provisions of this treaty. The legislative proposal is not intended to authorize any agency to prescribe substantive, generally applicable regulations. The regulations FAA would develop would simply outline the procedural aspects of getting a Mission Authorization, consistent with the case-by-case interagency process outlined above.

In addition to providing a regularized, predictable mechanism for authorizing commercial space activities, the Mission Authorization proposal is designed to preserve the competitiveness of the American launch industry. At present, United States Government review processes tied to the launch licensing framework—such as the Payload Review process—are limited to payloads launched from the United States. To the extent payload owners perceive these existing processes as presenting regulatory risk or inconvenience, they serve as a disincentive for purchasing launch services from American providers. By contrast, the authorization requirement in the Mission Authorization proposal would apply to United States nationals irrespective of launch location, thus enhancing the global competitiveness of the American launch industry.

The proposed Mission Authorization framework in the Appendix is not intended to affect existing space activities such as launch services, communications, or remote sensing for which current regulation by the FAA, FCC, or NOAA is sufficient to fulfill the United States' obligations under the Outer Space Treaty.

Sincerely,
John P. Holdren
Director and
Assistant to the President for Science and Technology
cc: Senator Bill Nelson
Representative Eddie Bernice Johnson
Senator Ted Cruz
Senator Gary Peters
Representative Brian Babin
Representative Donna Edwards

Mission Authorization Proposal

Chapter 509 of title 51, United States Code, is amended—

(a) In section 50902, by adding between subparagraphs (8) and (9) the following definition, "mission" means the operation of a space object, with or without human occupants, in outer space, including on the Moon and other celestial bodies.
(b) By inserting "mission," after "reentry site," in section 50919(g).

(c) By inserting after section 50923,
  a. Section 50924, Mission Authorization—(a) The Secretary of Transportation, in coordination with the Secretary of Defense, the Secretary of State, the Secretary of Commerce, the NASA Administrator, the Director of National Intelligence, and such other appropriate United States Government departments and agencies as the Secretary deems appropriate, is authorized to grant authorizations for missions in outer space. The Secretary shall grant such authorizations to the extent consistent with the international obligations, foreign policy and national security interests of the United States, and United States Government uses of outer space, with such conditions as the Secretary, in coordination with Secretary of Defense, the Secretary of State, the Secretary of Commerce, the NASA Administrator, the Director of National Intelligence, and other appropriate departments and agencies, deems necessary for compliance with United States international obligations, preservation of the foreign policy interests and national security of the United States, and protection of United States Government uses of outer space.
  1. No person that is subject to the jurisdiction or control of the United States may, directly or through any subsidiary or affiliate, conduct missions in outer space without authorization under this section.
  2. The following classes of Missions are exempt from this authorization requirement:
     i. Government activities subject to section 50919(g);
     ii. Missions for which licensing by the Department of Transportation under Chapter 509 of Title 51, the Federal Communications Commission under the Communications Act of 1934 (47 U.S.C. 151 et seq.), or by the Secretary of Commerce under chapter 601 of Title 51, is sufficient to fulfill the United States obligations under the Outer Space Treaty;
     iii. Missions, or aspects thereof, conducted for or with one or more United States Government departments or agencies, unless the Secretary and the relevant departments or agencies determine that an authorization is required to provide effective supervision of the mission, or aspects thereof;
  (b) MISSION AUTHORIZATION REGISTRY—The Secretary shall maintain a registry of Mission Authorizations and the information contained therein. The Secretary is authorized to require the holder of a Mission Authorization to provide updated information both on a periodic basis, and whenever the holder of the authorization experiences a material change to operations that would affect the affirmations and

information that were originally submitted in support of the authorization. In the event of such material changes in operations, the Secretary, in coordination with Secretary of Defense, the Secretary of State, Secretary of Commerce, the NASA Administrator, the Director of National Intelligence, and other appropriate departments and agencies, shall make such modifications to mission authorizations as necessary for compliance with United States international obligations, preservation of the foreign policy interests and national security of the United States, and protection of United States Government uses of outer space.

(d) By inserting at the end of chapter 509,

a. Section 50925, Conjunction Analysis—The Secretary of Transportation, in coordination with the Secretary of Defense, is authorized to examine the planned and actual operational trajectories of space objects and advise operators as appropriate to facilitate prevention of collisions.

Glossary of Key Terms and Phrases

Active space debris removal The various possible means by which space debris can be actively removed so as to prevent collisions that can create thousands of new debris elements. The increasing buildup of space debris in orbit, especially in low Earth orbit (LEO) and even in geosynchronous orbit (GEO), is of increasing concern.

Astral abundance The idea that there are vast and seemingly unlimited resources available in all the asteroids, comets, planets and moons in the Solar System and beyond. For instance, there is far more water in the billions of asteroids in the Solar System than in all the oceans on Earth.

Citizen astronauts This is a term coined by Sir Arthur Clarke to refer to citizens who can book passage on a commercial spaceplane and fly on a suborbital flight high enough, i.e. over 100 km high, so as to be considered astronauts.

Commercial Spaceflight Federation A "New Space" not for profit organization that seeks to advance the cause of new commercial space enterprises with a special emphasis on spaceplane development and commercial spaceflight for "citizen astronauts."

Cosmic hazards There are many types of cosmic hazards that should be considered a significant threat to modern humanity. These include a major asteroid or comet strike, with an asteroid of 35 m in diameter being a "major city killer," solar radiation flares, coronal mass ejections from the Sun, shifts in Earth's magnetic field that alter the Van Allen Belt's protective shielding of Earth, and runaway orbital space debris known as the Kessler Syndrome (see Planetary Defense).

Cubesat   A very small 1 unit cube-shaped satellite 10 cm × 10 cm × 10 cm that is often used for student experiments. There are larger versions that are 2–6 units that are used for many new applications for commercial applications.

Earth's finite resources   There have been various warnings about the world's mounting human population and the limits of natural resources, potable water, and food that the planet can sustain. Thomas Malthus was the first to publicly warn of such a concern, but in more recent times there has been the Club of Rome "Limits to Growth" study, the book Population Bomb and many other books and studies.

Global Navigation Satellite Services or Global Navigation Satellite System (GNSS)   Satellites used to provide navigation and targeting capabilities as well as precision timing. There are several such systems now in operation. These include the U. S. system known as GPS, the Russian system known as Glonass, the Japanese system known as Quazi-Zenith Satellite System, the Chinese systems known as Beidou and Compass, the European system known as Galileo, and the Indian system known as the Indian Regional Navigation Satellite System.

GPS   The Global Positioning Satellite system that is also known as NAVSTAR. This is the U. S. GNSS system that consists of 27 satellites in medium Earth orbit that is operated by the U. S. Department of Defense to support, among other functions, the targeting of missiles and so-called "smart bombs."

Launch vehicle or rocket   A system for launching satellites, experimental probes, and crewed spacecraft. More recently reusable rockets and spaceplanes have also been developed. These rocket systems typically use solid or liquid explosive fuels. Electronic systems powered by ions are used for thrusters to maneuver in space but are not currently able to launch from Earth's surface. Sounding rockets do not go into orbit but return to the ground. It requires sustained higher thrust to launch a satellite into higher and higher orbits. Ultimately so-called escape velocity would allow a payload to go beyond the reach of Earth's "gravity well" or beyond the reach of its gravitational pull.

Liability Convention of 1972   Formally known as the Convention on International Liability for Damage Caused by Space Objects, it places on all "launching nations" an "absolute liability" for any damage caused by a space object that it launches. The Russian satellite with a radioactive power plant that crashed into Canada would have come into play if Russia and Canada had not come to a negotiated settlement.

Megacities A city with a population of over ten million people. The number of such global cities is now in excess of 25 and will continue to increase over time.

Meteorological, or weather, satellites Satellites used to monitor and track weather patterns and storms and are also used to monitor "space weather" and solar storms and to track patterns of climate change.

Moon Agreement Known officially as the "Agreement Governing the Activities of States on the Moon and Other Celestial Bodies." It is controversial in that it has not been agreed to by the major spacefaring nations, and it took a number of years to acquire the signatures needed for it to come into force.

Multi-planet civilization Noted astrophysicists such as Carl Sagan and Stephen Hawking have indicated that a range of cosmic hazards threaten the long term survival of Homo sapiens as a sustainable species unless we are able to create viable habitats on different planets. They suggest that all of our eggs are literally in one basket and that survival requires moving beyond Earth.

"New Space" enterprise or economy The phrase "New Space" refers to the rise of new types of commercial space enterprises that are independent of the military or the government and are generally thought to involve start-up companies rather than well established aerospace companies.

On-orbit servicing The ability to dock with a space object to carry out some form of servicing. This could include refueling, upgrade, retrofit or repair of satellite and its components, repositioning of a satellite, or even harvesting of components such as antennas or solar arrays so that these parts could be redeployed for other uses.

Outer Space Treaty of 1965 The main and overarching space treaty on which four other agreements or conventions are based. It was developed by the U. N. Committee on the Peaceful Uses of Outer Affairs, agreed by the U. N. General Assembly and now ratified by well over 100 countries, including all major spacefaring nations.

Planetary defense Refers to the many cosmic dangers that could threaten the long-term survival of humanity. The five mass extinction events that have occurred over time can be attributed to overheating by the Sun or the impact of a giant 5-km-wide asteroid some 65 million years ago with the so-called K-T event. This event is thought to have ended the life of some 75 % of all plant and animal life on the planet.

Planetary migration The migration of humanity to another planet that has been "terraformed" to be livable for humans, animals and plant life on a sustained basis.

Population growth The increase of worldwide human population, which has grown from 800 million in 1800 to 1.8 billion in 1900 to 7.1 billion in 2000 and is projected to grow as high as 12 billion by 2100, although some projections see the possibility of decline in the latter half of the twenty-first century with a peak between 9 and 10 billion.

Protozone Area above Earth's orbit but below outer space where it is expected that much development will take place in future years but for which there is almost no legislation to govern such development.

Registration Convention of 1975 Formally known as the Convention on Registration of Objects Launched into Outer Space. It places a formal requirement on all nations to register all objects launched into outer space. Under the Liability Convention, all "launching nations" are absolutely responsible for all objects they launch into outer space.

Remote sensing satellite A satellites that is used to find resources, such as in the ocean or outer space, and other information using satellite-based sensors that include radar, optical sensing, ultraviolent and infrared sensors. The latest innovation in this field involves what is called "hyperspectral sensing," which allows data to be collected in narrow frequency bands or well defined spectral areas to determine much more precise information about the remotely sensed areas. These satellites also have higher specral resolution.

Satellite communications The most important of the commercial uses of applications satellites to provide telecommunications, mobile satellite communications, data networking, and radio and television broadcasting. The commercial size of this market is estimated to be in excess of $150 billion/year. There are, in addition, extensive military communications systems that provide tactical and non-tactical communications for a number of countries.

Satellite constellations A configuration of multiple satellites that are deployed over Earth in some sort of pattern or even random deployment so as to provide planetary coverage by a network of small satellites. Examples of these include Iridium and Globalstar for mobile satellite communications, Terra Bella and Planet Labs for remote sensing, and the planned OneWeb satellite constellation that is optimized for low latency Internet-based networking. All these systems and more are in low Earth orbit.

Small satellites General term that refers to extremely small satellites weighing only a few grams or ounces, ranging from cubesats that might be in the 5–20 kg (11–44 pound) to quite capable commercial small satellites in constellations (i.e., twenty to even many hundreds of satellites blanketing Earth that would typically be in low Earth orbit) to provide communications, remote sensing or other services.

Space billionaires This refers to key people that have started major "New Space" ventures of note, including Paul Allen, Jeff Bezos, Robert Bigelow, Richard Branson, and Elon Musk.

Space commercialization Refers to the shift of space-related activities from government to commercial enterprises funded by private investors and entrepreneurs.

Space debris Generally considered to refer to all defunct, inoperable or inert objects in space without independent means of control. It is estimated that there are now over 6000 metric tons of such debris, of which about 40 % is concentrated in low Earth orbit and especially in polar orbit or objects that travel over or near to the Earth's North and South Poles. Also referred to as space junk.

Space elevator This is a concept of using a cable or tether as a lifting system to connect Earth and the geosynchronous, or "Clarke," orbit or alternatively to connect Earth and the Moon. The idea is that this would be much more energy and environmentally efficient and much lower cost than the use of chemical rockets to move materials into space. Such a system could also be used to move materials from space to Earth.

Space entrepreneurs The ambitious individuals who have the drive, new ideas, and supporting investment funds needed to launch new space ventures.

Space ethicist A philosopher who devotes time, energy and research into the areas of interest and concern related to the use, exploitation, research into, and future ways forward with regard to outer space and especially human and human-directed activities above the atmosphere of our planet.

Space mining Proposed future activity that would involve the extraction of resources from objects in outer space. Currently there are many unanswered questions about this activity. How large is something to be called a "celestial body" and is it legally acceptable to remove material from a celestial body?

Space navigation, tracking and precision timing Refers to the ability to use satellites with precision timing from atomic clocks to be able to carry out extremely accurate geo-location measurements. The timing from these satellite-based clocks can allow synchronization of data networks (i.e., the Internet), time stamping of banking and stock transactions, etc. (see also GPS).

Space material processing and manufacturing The logical next step beyond space mining or material processing, which involves the processing of these materials so that they might be usefully applied to manufacturing satellites or producing rocket fuel. All of the companies now seeking to engage in space mining have also indicated that they would also want to do space processing and manufacturing as well.

Spaceship Earth   Planet Earth is for humanity simply a 6 sextillion- ton spaceship that currently carries a very large crew of about 7.5 billion people and a lot of other flora and fauna as well. If Earth were a cosmic apple, the protective shielding we call the atmosphere is equivalent in size to the rind of the apple. The Van Allen Belts formed by Earth's magnetic poles is all that protects us from destruction from periodic coronal mass ejections from the Sun.

U. N. Committee on the Peaceful Uses of Outer Space (COPUOS)   A committee that meets once a year in Vienna. It has a Technical Subcommittee and a Legal Subcommittee that both meet once a year. In addition it has Working Groups. Currently the Working Group on the Long-Term Sustainability of Outer Space Activities is seeking to get broad consensus agreement on a series of steps that could be taken to address the long term sustainability of space that would address such areas as space weather, orbital space debris, etc. This Committee now has over 80 members.

U. N. Office of Outer Space Affairs (OOSA)   This is the office headquartered in Vienna, Austria that serves as the secretariat for the UN Committee on the Peaceful Uses of Outer Space. Its various activities are listed on its website.

Urban density   Refers mainly to the percentage of people worldwide that live in towns and cities (now around 53 %) versus those that live in rural areas (currently 47 %). This urban density is estimated to grow to over 80 % and perhaps 85 % by the end of the twenty-first century.

U. S. Commercial Space Launch Competitiveness Act of 2015   Legislation that was enacted in late 2015 that covered a number of different topics to encourage commercial launch development. Title 4 of this act created legislation to encourage the development of commercial space mining activities. This act is somewhat controversial in that some consider the provisions related to space mining to be contrary to the provisions of the Outer Space Treaty and the Moon Agreement.

Index

A
Active space debris removal, 175
"Africastar" satellite, 48
AI computer programs, 24
Arkyd space telescope, 13, 14
Artificial intelligence (AI), 2, 6, 11–13, 24–26, 35, 37, 194, 196, 199, 206, 207, 209, 211
"Artificial Van Allen Belts", 198
Asteroid Belt, 31
Astral abundance
    Arkyd space telescope, 13, 14
    artificially intelligent, 15
    Disrupt Space, 12
    energetic enterprise, 14
    fusion-based power plants, 15
    non-renewable resources, 10
    singularity, 12
Automatons, 24

B
B2B. *See* Business to business (B2B) satellite services
B330, 145, 146, 148
BEAM. *See* Bigelow expandable activity module (BEAM)
Bigelow Aerospace, 6, 8, 11, 69, 141, 144–148
Bigelow expandable activity module (BEAM), 145, 148
Biosphere 2, 151–152
Biospherians, 23, 151, 153
Business to business (B2B) satellite services
    LEO constellations, 56
    lower VHF and UHF frequency bands, 56
    MEO and GEO, 56
    TDRS system, 56

C
Cable-based electronic entertainment, 21
CANR. *See* Chemically assisted nuclear reaction (CANR)
Chemically assisted nuclear reaction (CANR), 96
Chinese National Space Agency, 173, 204
Civilian satellite, 112
Clarke orbit, 86, 87

Climate change, 23, 28, 30, 36, 60, 62–64, 75, 92, 110, 124, 127–138, 153, 192, 194, 195, 200, 202, 205, 206, 209
Comets
 K-T mass extinction, 130, 131
 NEOCam, 133
 NEOWISE, 132
 warning systems, 132
Commercial off-the-self (COTS), 44
Commercial satellite, 112
Commercial space enterprises, 11
Commercial space transport, 71
 challenge for space agencies, 87
 3D printers, 71
 economic and technical challenges, 87–89
 hypersonic jets, 86
 legal challenges, 87–89
 low cost launchers, 82–83
 on-orbit robotics, 71
 protospace, 84–86
 regulatory challenges, 87–89
 space elevators, 76
 space tourism, 69, 76
 spaceplanes, 76
 SpaceX (*see* Space exploration technologies corporation (SpaceX))
 Stratolaunch, 72
Commercial transportation systems, 15
Communications satellite, 4, 37, 40–42, 52, 53, 55, 60, 82, 94, 103, 109, 111, 113, 119, 136, 150, 169 (*see also* Military and governmental communications satellites)
Communications satellite industry
 business entities, 46
 categories of, 46
 DABS, 47–48
Conventional communications satellite systems, 40

COPUOS. *See* U.N. Committee on the Peaceful Uses of Outer Space (COPUOS)
Coronal mass ejections, 62, 65, 92, 98, 123, 129, 130, 134, 135, 138, 151, 152, 163, 173, 192, 204
Cosmic hazards, 4, 92, 99, 173, 203, 204
 asteroids and comets, 192
 debris, 128
 defense, 122–123
 GPS Navstar system, 128
 planetary defense, 138
 protection, 191, 197
 risk, 128–130
 solar storms, 134–136
 space debris, 136–137
COTS. *See* Commercial off-the-self (COTS)

D

DABS. *See* Direct audio broadcasting service (DABS)
Data relay. *See* Business to business (B2B) satellite services
Deep Space Industries (DSI), 100, 169
Defense
 communications, 112
 cosmic hazards, 122–123
 defense-related space systems, 88–89
 space expenditures, 111
Direct audio broadcasting service (DABS)
 "Africastar", 48
 offerings, 47
 OnStar satellite service, 48
 XM/Sirius, 47
Disrupt Space, 2, 12
3D printing technology, 96
Dreamchaser spaceplane, 70
DSI. *See* Deep Space Industries (DSI)

E

Earth's finite resources
  overuse, 192
  recycling, 202
  sustainability practices, 192
Echostar XVII satellite, 42, 43
Electromagnetic pulses (EMPs), 111, 134
EMPs. *See* Electromagnetic pulses (EMPs)
Environmental and health issues, 161, 172
Eric Anderson envisions, 22
ESA. *See* European Space Agency (ESA)
e-Sphere, 22
ETSI. *See* European Telecommunications Standards Institute (ETSI)
European Space Agency (ESA), 52, 61, 70, 102, 130, 204
European Telecommunications Standards Institute (ETSI), 67

F

Fixed satellite services (FSS), 46
  commercial system, 48
  "Early Bird", 48
  high throughput satellites, 50
  highest capacity, 49
  Moore's Law, 49
  national and international regulatory authorities, 50
  satellites *vs.* fiber optic networks, 51
  TASI, 48
  types of services, 49
  VSATs and USATs, 49

G

GeoEye, 28, 29, 61
Geographical Information Systems (GIS), 63
GIG. *See* Global Information Grid (GIG)
GIS. *See* Geographical Information Systems (GIS)
Global commons. *See* Outer Space Treaty
Global Governance of Outer Space, 36, 170–176
Global Information Grid (GIG), 58, 113
Global Navigational Satellite Systems (GNSS), 63, 170, 171
Global Positioning Satellite (GPS) system, 62–64, 120, 128, 135
Global society, 200–211
  Earth's magnetic field, 191
  global agreement, 198
  global governance, 196
  long-term human survival, 194
  long-term plan, sustainable growth, 194
  long-term survival, 197
  magnetic space shields, 195
  natural and cosmic threats, 195
  "natural laws" of science, 191
  population growth, 192, 195
  protection, cosmic hazards, 191
  resources, skies, 197
  space dividends, 199
  sustainable practices, 192
  sustainment of life, 192
  Ten Point Program
    (*see* Ten Point Program)
Global space security, 9, 172
GLONASS, Soviet/Russian system, 63
GNSS. *See* Global Navigational Satellite Systems (GNSS)
Governance and control systems, 185, 186
GPS. *See* Global Positioning Satellite (GPS) system
Gravity assist, 87

H

Habitats
  B330, 145, 146
  BEAM, 145
  Bigelow aerospace, 146, 147
  LaGrange point, 148–149
  long-term tourism, 148
  lower cost and short term tourism, 147
  Mars, 148–149
  Moon, 148–149
  NASA, 148
Halo optical satellite system, 59
HAPS. See High altitude platform system (HAPS)
High altitude platform system (HAPS), 85, 89, 171
High throughput satellites, 40, 42, 43, 49, 50, 55, 66
House of Representatives bill HR2262, 164
Human space missions, 173
Hypersonic spaceplane, 117, 118

I

IAASS. See International Association for the Advancement of Space Safety (IAASS)
IADC. See The Inter-Agency Space Debris Coordination Committee (IADC)
ICG. See International Committee on Global Navigation (ICG)
IEC. See International Electro-Technical Commission (IEC)
IEEE. See Institute of Electrical and Electronics Engineers (IEEE)
IETF. See Internet Engineering Task Force (IETF)
Indian Regional Navigational Satellite System, 63
Indian Space Research Organization (ISRO), 204
Institute of Electrical and Electronics Engineers (IEEE), 67
Intelsat's Epic Satellites, 42
The Inter-Agency space Debris Coordinating (IADC), 67, 163, 176, 187
International Association for the Advancement of Space Safety (IAASS), 83, 173, 174, 176, 187
International Asteroid Warning Network (IAWN), 138, 187, 204
International Civil Aviation Organization (ICAO), 81, 89, 166, 167, 171, 173, 187
International Committee on Global Navigation (ICG), 64, 67
International Electro-Technical Commission (IEC), 67
International Space University, 12, 36, 71, 76, 82, 199
International Standards Organization (ISO), 67
International Telecommunication Union (ITU), 55, 56, 67, 118, 119, 166, 171, 187
Internet Engineering Task Force (IETF), 67
Internet entertainment, 21
Inter-orbital Systems (IOS), 147, 148
IOS. See Inter-orbital Systems (IOS)
ISO. See International Standards Organization (ISO)
ISS national laboratory, 214
ITU. See International Telecommunication Union (ITU)

J

Japanese Space Agency (JAXA), 56, 61, 80, 81, 118, 204
JAXA. See Japanese Space Agency (JAXA)

K

K-T mass extinction, 131

L

L-1 Lagrange point, 114
LaGrange point, 148–151, 188, 205
Launch services, 214
Launch vehicles, 74, 78, 82, 144, 182
    Soyuz, 78
    SpaceXFalcon 9, 74
    Strato, 45
LENR. *See* Low energy nuclear reaction (LENR)
The Liability Agreement, 170
Liability Convention, 104, 105, 116, 161
Low energy nuclear reaction (LENR), 96

M

Machine intelligence, 25, 106, 153
Mars
    habitats, 148–150
    space colonies, 149
Mass driver device, 150
Mega cities, 12, 201, 204, 206
Mega LEO OneWeb system, 44
Meteorological satellites, 5, 34, 37, 171
    low Earth Sun synchronous polar orbiting satellites, 64
    NOAA, 65
    spacefaring nations, 64
    storms, 65
Military and governmental communications satellites
    dual-use purposes, 57
    GIG, 58
    hybrid systems, 57
    innovative XTAR satellite, 58
    UAV imaging, 57
Military uses, protozone and earth orbit, 167
Millimeter wave bands, 50
Mission Authorization Proposal, 217–220
Mobile satellite services (MSS), 46
    description, 51
    FSS, 54
    handsets, 55
    INMARSAT organization, 52
    ITU, 55
    large aperture antenna design, 54
    Marisat satellite, 52
    offerings, 54
    practical and economic consequence, 52
Moon
    habitats, 148–150
    space colonies, 149
Moon Agreement, 104, 165, 169, 174, 179, 180, 182, 210
The Moon Treaty, 164, 199

N

NASA, 199, 203
    cosmic hazard detection, 204
    long-term space habitats, 148
    WISE Infrared Telescope, 204
National Aeronautics and Space Administration Authorization Act of 2010, 214
National security, 111–113, 115
Nautilus space complex module. *See* B330
Navigation and timing satellites, 5
NEA. *See* Near Earth asteroid (NEA)
Near Earth asteroids (NEAs), 76, 100
Near earth objects (NEOs), 27, 31
NEAs. *See* Near Earth asteroids (NEAs)
NEOCam, 133
NEOs. *See* Near earth objects (NEOs)
NEOWISE, 132
New Space commerce, 35, 37

"New Space" enterprise/economy, 21, 24, 26, 33
  agreements, 17
  allocation, radio frequencies, 168
  astral abundance, 10–11
  billionaires, 7–8
  commercial space enterprises, 11
  commercial transportation systems, 15, 16
  energy, 17
  energy systems, 16
  entrepreneurs, 11
  frontier, 8–10
  Global Governance of Outer Space conference, 170
  habitats/facilities, 17
  homo sapiens, 7
  House of Representatives bill HR2262, 164
  hypersonic transport, 17
  issues, 160
  jobs, 16
  military uses, protozone and earth, 167
  outer space mining, 169
  private entities, 160
  public discontent, 159–160
  resources, 16
  solar power satellite systems, 168
  space debris removal, 167, 168
  space mining, 17
  space treaties, 161–163
  standards or regulatory actions recommended, 170–175
  technology, 18
  traffic management and control, 165, 166
  U. S. Commercial Space Launch Competitiveness Act of 2015, 165
New Space technology, 25
NOAA. *See* U.S. National Ocean and Atmospheric Administration (NOAA)
Non-renewable resources, 10
Noosphere, 35

O

Office of Science and Technology Policy, 213
OneWeb 700 satellite, 44
On-orbit servicing, 71, 88, 147, 171
OOSA. *See* U. N. Office of Outer Space Affairs (OOSA)
Operational satellite networks, 47–56
  airline operations, 41
  breakdown of, by function, 39, 40
  conventional communications, 40
  COTS, 44
  DABS (*see* Direct audio broadcasting service (DABS))
  disruptive technologies, 45
  electric motors and solid state digital processors, 41
  FSS (*see* Fixed satellite services (FSS))
  global infrastructure, 39
  high throughput satellites, 40
  Intelsat's Epic Satellites, 42
  MSS (*see* Mobile satellite services (MSS))
  networking constellations, 45
  next generation air traffic management, 41
  satellite navigation, 41
  SIA reporting system, 39
  Skybox, 45
  strategic and social benefit valuation, 40
  television broadcasting and home satellite services, 46, 47
  Terracom, fiber optic cable, 44
  WorldVu company, 42
Outer space security, 115–116
Outer Space Treaty (OST), 13, 104, 105, 122

COPUOS meeting, 162, 163
exploration, 178, 179
IADC, 163
international agreements, 187
international space law, 178
Liability Convention, 161
limitations, 170
mechanisms, 187
Moon Agreement, 179
space traffic management, 161
space weapons deployment, 167

P

Pacific International Space Center for Exploration Systems (PISCES), 149
Paul Allen envisions, 22
Peter Diamandis envisions, 22
PISCES. *See* Pacific International Space Center for Exploration Systems (PISCES)
Planetary colonization, 174
Planetary defense, 34, 75, 173, 175, 194, 198, 204, 209
  action paln, 138
  cosmic hazards, 122–123
  programs, 137
Planetary migration, 174
PNT. *See* Precision navigation and timing (PNT) satellites
Population growth, 12, 36, 91, 156, 192
  minimization, China, 202
  reductions, 202
Precision navigation and timing (PNT) satellites, 63
Private Missions Beyond Earth's Orbit, 215
Private space station, 171
Protozone, 33, 84–86, 118
  aircraft and jets, 113
  applications, 117
  commercial airspace, 116

hypersonic spaceplane, 118
radio frequencies (*see* Radio frequency)
traffic control and management issues, 117
uses, 117

R

Radio frequency
  Earth orbit, 118, 119
  geostationary orbit, 120
  ITU, 118, 119
  protozone, 118, 120
  satellite communications systems, 118
Ray Kurzweil envisions, 22
Remote sensing, 214
Remote sensing satellite, 5, 44, 62, 63, 67, 120, 171
  commercial operator GeoEye, 28
  RADARSAT 1 and 2, 61
Remote-sensing industry
  communications satellites, 60
  3-D hyper spectral data cube, 61
  infrared, optical and ultraviolet sensors, 60
  meteorological satellites, 62
  shutter controls, 62
  smart farming, 60
  space agencies, 61
  unusual purposes, 60
Robert Bigelow envisions, 22
Rocket launchers, 197
Rods from Gods, 111
RosCosmos. *See* Russian Space Agency (RosCosmos)
Russian Space Agency (RosCosmos), 61, 78, 173, 204

S

Saharan Desert, 29
Satellite broadcasting, 171

Satellite communication, 17, 95, 103, 118, 168, 214
　established space businesses, 34
　space communications, 26
Satellite constellation, 82, 118, 171
Satellite earth imaging, 27
Satellite entertainment, 21
Satellite Industry Association (SIA), 39
Satellite navigation and timing systems
　Clinton administration, 64
　geomatics, geospatial analysis, GIS, 63
　GLONASS, 63
　GPS system, 62
　ICG, 64
　PNT and GNSS, 63
　weapon systems, 62
Satellite telecommunications, 171
S-band radar system, 110
Security
　outer space, 115, 116
　protozone, 116–117
　space mining, 121–122
Sentinel infrared telescope, 183, 188
Shackleton Energy, 11, 99, 100, 103, 169, 180
SIA. *See* Satellite Industry Association (SIA)
Singularity. *See* Astral abundance
The Singularity Is Near, 25
The Singularity University, 199
Small satellites, 42, 43, 56, 70, 82, 83, 118
Smart digital processors, 27
Smart robots, 24, 105
Smart systems, 24
Solar power satellite systems, 34, 168, 169
　challenges, 95–97
　electromagnetic frequency, 103
　international standards and controls, 106
　laser/ radio frequency transmission, 92
　legal and regulatory framework, 103

　massive solar storms, 98
　P/V systems, 93
　photovoltaic cells, 92, 93
　rectenna, 94, 95
　RF/laser transmitter systems, 93
　solar shield, 99
　Van Allen Belts, 98
　wood-burning systems, 97
Solar System, 6, 13, 16, 31
Space abundance, 22
Space adventures, 34
Space age innovators, 21
Space billionaire, 7, 8, 43, 45, 71, 72, 93, 160
Space colonies, 24, 34, 192, 195, 197
　catcher system, 151
　genetic seeds, 155
　LaGrange point, 149
　mass driver, 150
　off-world economy, 143
　population control, 154–156
　self-sufficiency and long-term viability, 142
　solar storms, 153
　temperature gradients, 153
　terraforming operation, 154
Space commercialization, 162, 180
Space debris, 136, 137
　IADC, 163
　mitigation and removal, 162, 167, 168
　remediation and environmental oversight, 88
Space elevators/mag-lev systems, 17, 76, 86, 144
Space enterprises, 24, 33–34
Space entrepreneurs, 11, 21, 78, 96, 143, 144, 159, 197
Space Exploration Technologies Corporation (SpaceX)
　Falcon 9 Heavy vehicle, 74
　launcher development, 83
　Rutherford engine, 84
Space governance, 164, 165, 167–170

Global Governance of Outer Space
conference, 170
international regulation, 160
issues, 160
legal interpretations, space treaties,
161–163
model laws
allocation, radio frequencies, 168
House of Representatives bill
HR2262, 164
military uses, 167
mining and space resources,
169, 170
solar power satellite
transmissions, 168, 169
space debris mitigation and
removal, 167
space traffic management and
control, 165, 167
U. S. Commercial Space Launch
Competitiveness Act, 2015,
165
public discontent, 159, 160
standards/regulatory actions,
recommended, 170–175
Space habitats, 34
Space hazards, 173
Space junk, 101, 130, 136, 204
Space launch services, 173
Space materials processing and
manufacturing, 192
Space mining, 13, 17, 19, 30, 32, 33,
36, 65, 170, 192, 197
"abiotic" resources, 103
Akyrd-3R probe, 101
asteroids, 164, 169
DSI, 100
helium-3 isotopes, 101
international standards and
controls, 106
legal status, 103
licensed authorization, 104
moon, 169

NEA, 100
Planetary Resources, 99
safety and legal liability, 105
Shackleton Energy, 100
smart robots, 106
U. S. Commercial Space Launch
Competitiveness Act of 2015,
169
U. S. regulatory actions, 169
Space Mission Planning Advisory
Group (SMPAG), 204
Space navigation, 17, 34, 37, 38, 66
Space R&D programs, 33
Space Resource Exploration and
Utilization Act of 2015
common heritage of mankind, 181
global commons, 182
legal enforcement, 182
outer space change, 181
policing, 182, 184
regulatory system, 184
Sentinel infrared telescope, 183
space colonies, 181
traffic control and management, 183
Space Resource Utilization, 215
Space Swiss Systems (S3), 70
Space tourism, 73, 75, 76, 78, 80, 81
Space transportation, 15, 16, 22, 37, 188
ICAO, 166
radiation danger, 166
radio frequencies, allocation of, 166
SARPS, 166
traffic management and control,
166, 172
Space-based navigation, 167
Space-based war-fighting systems, 109
SpaceHab, 144, 145
Spaceplane system, 34
aerospace organizations, 79
safe and non-polluting, 80
development, 25
Space Ship 2, 69
Space Swiss Systems (S3), 70

Spaceplane system (*cont.*)
  SpaceShipOne, 77
  and space tourism, 77
SpaceShipOne, 77
SpaceShipTwo, 79
Star wars, 109
Stratobus, 85
Stratolaunch, 72
Subspace/protospace, 9
Super automation, 26
Super urbanization, 194, 200, 201, 205
Syncom 2, 4

T

TASI. *See* Time Assignment Speech Interpolation (TASI)
TDRS. *See* Tracking and Data Relay Satellite (TDRS) system
Ten Point Program
  the global commons, 208–210
  global population control, 202
  humanity, 210
  laws and regulation, 208
  mega-structures and intellectual infrastructure, 205
  planetary protection programs, 203–205
  singularity, 207, 208
  space- and ground-based infrastructure, 206, 207
  sustainability, 203
  urban sprawl, 205, 206
Time and human technological progress, 24
Time Assignment Speech Interpolation (TASI), 48
Tiny 40-kg Early Bird satellite, 43
Tracking and Data Relay Satellite (TDRS) system, 56
Transformational Satellite System (TSat), 59
Transitional satellite (TSAT), 110
Treaty on Principles Governing the Activities of States in the Exploration and Use of Outer Space, 215
Tripartite space governance unit, 185
TSat. *See* Transformational Satellite System (TSat)
TSAT. *See* Transitional satellite (TSAT)

U

U. N. Committee on the Peaceful Uses of Outer Space (COPUOS), 18, 67, 116, 187
  consensus, 162
  membership, 162
  sustainability of space, 162
U. N. Office of Disarmament Affairs (UNODA), 116
U. N. Office of Outer Space Affairs (OOSA), 64, 178, 187
U. S. Commercial Space Launch Competitiveness Act of 2015, 169
U. S. Commercial Space Launch Competitiveness Act, Public Law 114-90, 213–220
U. S. National Ocean and Atmospheric Administration (NOAA), 65
UIMA. *See* Unstructured Information Management Architecture (UIMA)
Ultra-small aperture terminals (USATs), 49
UNISPACE + 50 Conference, 2018, 176
United Kingdom's Outer Space Act of 1986, 216
UNODA. *See* U. N. Office of Disarmament Affairs (UNODA)

Unstructured Information Management Architecture (UIMA), 26
Urban density, 201, 205, 206
USATs. *See* Ultra-small aperture terminals (USATs)

V

Van Allen Belts, 98
VCLS. *See* Venture Class Launch Services (VCLS)
Venture Class Launch Services (VCLS), 82, 84
Very small aperture terminals (VSATs), 49
Viasat 2, 4
Vomit comet, 79
VSATs. *See* Very small aperture terminals (VSATs)

W

Warfare and national protective systems, 112–113
Weapons systems
 commercial satellite, 112
 defensive and offensive, 110
 space military systems, 116
Weather satellites, 41, 109, 152

MIX
Papier aus verantwortungsvollen Quellen
Paper from responsible sources
FSC® C105338

If you have any concerns about our products,
you can contact us on
ProductSafety@springernature.com

In case Publisher is established outside the EU,
the EU authorized representative is:
**Springer Nature Customer Service Center GmbH
Europaplatz 3, 69115 Heidelberg, Germany**

Printed by Libri Plureos GmbH
in Hamburg, Germany